SpringerBriefs in Computer Science

Series editors

Stan Zdonik
Shashi Shekhar
Jonathan Katz
Xindong Wu
Lakhmi C. Jain
David Padua
Xuemin (Sherman) Shen
Borko Furht
V.S. Subrahmanian
Martial Hebert
Katsushi Ikeuchi
Bruno Siciliano
Sushil Jajodia
Newton Lee

More information about this series at http://www.springer.com/series/10028

Zdravko Galić

Spatio-Temporal Data Streams

 Springer

Zdravko Galić
Faculty of Electrical Engineering and Computing
University of Zagreb
Zagreb
Croatia

ISSN 2191-5768 ISSN 2191-5776 (electronic)
SpringerBriefs in Computer Science
ISBN 978-1-4939-6573-1 ISBN 978-1-4939-6575-5 (eBook)
DOI 10.1007/978-1-4939-6575-5

Library of Congress Control Number: 2016947747

Printed on acid-free paper

This Springer imprint is published by Springer Nature
The registered company is Springer Science+Business Media LLC New York

In loving memory of my mother Pavica (Lončar) Galić for all her love, care and support.

Preface

Recent advances in the information and communication technologies, especially the rapid development of wireless communication, sensor networks, mobile computing technologies, global navigational satellite systems (GPS, GLONASS, COMPASS, Galileo), RFID, wireless sensor networks and spatially enabled devices are leading to an exponential growth in the amount of available data produced continuously at hight speed. Due to the advancements in recent years, a new class of applications has come to the forefront: sensor networks, moving objects tracking, homeland security, fleet management, real-time intelligent transportation systems, etc. Applications in these novel domains process huge volumes of continuous streaming data, i.e. data that is produced incrementally over time, rather than being available fully before processing. According to the type of processing, data stream processing could be broadly classified into two categories: *data stream management* and *data stream mining*. Data stream management systems (DSMS) have been developed to querying and summarization of continuous data streams for further processing. Usually based on pure relational paradigm, they have rudimentary spatio-temporal capabilities.

An orthogonal issue to data stream management is data stream mining and knowledge discovery. Spatial and spatio-temporal knowledge discovery from data streams (KDDS) is based on premise that information is hidden in data streams in form of interesting spatial and spatio-temporal *patterns*. Among many kinds of spatio-temporal data, moving objects data are especially interesting and important, because they perfectly fit with the data stream concepts. Mining movement patterns of multiple moving objects is a motivating and challenging topic. These patterns include, but are not limited to, moving clusters, trajectory patterns, periodic patterns, group movement patterns (flock, convoy), etc. Knowledge discovery is a set of several processes, of which data mining is merely a crucial one. For that reason, our focus is limited to data mining algorithms, particularly to redesigning and adapting traditional clustering data mining algorithms for spatio-temporal data streams.

This book is an overview of the young, fast-growing and emerging fields of *spatio-temporal data stream* processing: management and data mining, and knowledge discovery. It evolved from the class notes of a postgraduate course "Spatio-Temporal Databases and Data Streams" at the University of Zagreb. Researchers and students both within and outside the Faculty of Electrical Engineering and Computing found the course very interesting and useful in their work. The book provides an introduction to the spatio-temporal stream processing, presents fundamental concepts and discusses design of prototypes to make the book useful for a diverse audience. Emphasis on the conceptual framework reflects the organization of the book.

The book can be used for teaching graduate or advanced undergraduate students, since it provides clear and concise presentation of major concepts and results in the field. This is a book also for computer science researchers and researchers from other disciplines and application, such as complex event processing, intelligent transportation engineering, telecommunication and environment, who desire to obtain an overview on the emerging spatio-temporal data streams field. It would be of interest to computer professionals, software developers and domain experts from industry (GIS experts, spatial data analysts, data scientists, real-time and big data analysts, BI analysts). Furthermore, experts in these disciplines could profit from the spatio-temporal data stream vision when designing and building innovative systems.

This book should be accessible to anyone with a general background on the concepts of weather database systems or complex event processing. A deeper knowledge of the spatio-temporal database systems and big data technologies would be an advantage. A background on data stream processing is useful but not needed—brief introduction is provided in Chap. 2.

I hope this book provides a useful overview of the young and evolving field of spatio-temporal data streams.

Zagreb, Croatia and Vienna, Austria   Zdravko Galić
June 2016

Acknowledgements

First of all, I would like to stress that this book would not have been possible without the original contributions of many researchers in the field of data streams.

The work described in this book has been supported by the grants from the Ministry of Science the Republic of Croatia under grant number 036-0361983-2010 ("Geospatial Sensors and Moving Objects Databases", Jan. 2007–Dec. 2012).

Writing this book is, in a way, a result of my research work on spatio-temporal data streams. I would like to express my gratitude to the following people for their help and encouragement in writing of this book. I am most grateful to my colleague Mirta Baranović, my postgraduate students, especially my PhD students Krešimir Križanović and Emir Mešković, as well as Dario Osmanović, with whom I had the pleasure to cooperate and work.

Special thanks to Susan Lagerstrom-Fife, Springer Computer Science editor and Jennifer Malat, Assistant Editor at Springer Science+Business Media, for their enthusiasm, patience and timely support throughout the preparation of this book. I am indebted to all of the reviewers for their invaluable feedback.

I am also truly grateful to Branka Marijanović for proofreading the draft book.

A word of gratitude and thanks to my M., family and close friends, who have been the major source of encouragement and support.

Contents

Acronyms

ACID	Atomicity, Consistency, Isolation, Durability
API	Application Programming Interface
BI	Business Intelligence
CEP	Complex Event Processing
CQ	Continuous Query
CQL	Continuous Query Language
DAG	Directed Acyclic Graph
DB	Database
DBMS	Database Management System
DMQL	Data Mining Query Language
DSMS	Data Stream Management System
DW	Data Warehouse
EPSG	European Petroleum Survey Group
FIFE	First-In-First-Expire
GIS	Geoinformation System
GKD	Geographical Knowledge Discovery
GNSS	Global Navigation Satellite System
GPS	Global Positioning System
HDFS	Hadoop Distributed File System
IFP	Information Flow Processing
IT	Information Technology
JVM	Java Virtual Machine
KDD	Knowledge Discovery in Databases
KDDS	Knowledge Discovery in Data Streams
LBS	Location-Based Services
ML	Machine Learning
MOD	Moving Objects Database
OGC	Open Geospatial Consortium
OLAP	Online Analytical Processing
OR	Object-Relational

PMML	Predictive Model Markup Language
QoS	Quality of Service
RFID	Radio Frequency Identification
RTCS	Real-Time Computing Systems
SPE	Stream Processing Engine
SQL	Structured Query Language
SRID	Spatial Reference System Identifier
stKDDS	Spatio-Temporal Knowledge Discovery in Data Streams
TCP	Transmission Control Protocol
UDA	User Defined Aggregate
UDF	User Defined Function
UDO	User Defined Operator
UDP	User Datagram Protocol
WGS	World Geodetic System
YARN	Yet Another Resource Negotiator

Chapter 1
Introduction

Abstract Information flow processing applications need to process a huge volume of continuous *data streams*. They are pushing traditional database, data warehousing and data mining technologies beyond their limits due to their massively increasing data volumes and demands for real-time processing. This chapter gives an overview of query processing in data stream management systems (DSMS) and the most influential academic prototypes, as well as available commercial products. Furthermore, this chapter also provides an outline of the basic concepts of spatio-temporal knowledge discovery from data streams, including a list of relevant data stream mining academic prototypes and commercial products.

Keywords Data streams · Data stream management systems · Spatio-temporal data streams · Data stream mining · Knowledge discovery

1.1 From Databases to Data Streams

Traditional database management systems (DBMSs) have been researched and used for a wide range of applications for over three decades. They have been used as a simple but effective warehouse of business data in applications that require persistent data storage and complex querying. A database consists of a finite, persistent set of objects that are relatively static, with insertions, deletions and updates occurring less frequently than queries. Queries expressed in a query language such as SQL are executed when posed and the answers reflect the current state of the database. Over the years, it has become obvious that many applications involving spatial data need extended and specialized DBMS functionalities. Spatial databases [17, 20, 51] evolved from traditional DBMS, and have been successfully implemented as an extension of DBMS based on object-oriented or object-relational paradigm [5, 44, 46]. More recently, spatio-temporal databases [36, 63] and their specific subclass called moving object databases (MOD) [19, 28, 40] have been an active area of research, with a few available research prototypes [6, 33]. However, DBMSs have proven to be well-suited to the organization, storage and retrieval of finite, persistent data sets including spatial as well as spatio-temporal ones.

© The Author(s) 2016
Z. Galić, *Spatio-Temporal Data Streams*, SpringerBriefs
in Computer Science, DOI 10.1007/978-1-4939-6575-5_1

Recent advances in the information and communication technologies, especially the rapid development of wireless communication, sensor networks, mobile computing technologies, global navigational satellite systems (GNSS), radio-frequency identification (RFID), wireless sensor networks and spatially enabled devices are leading to an exponential growth in the amount of available data produced continuously at hight speed. Due to these advancements, a new class of *monitoring applications* has come to the focus: sensor networks for monitoring physical or environmental conditions, moving objects tracking, real-time intelligent transportation systems, homeland security, network and infrastructure monitoring, etc.

These new information flow processing (IFP) application domains need to process a huge volume of continuous *data streams*, i.e. real-time, transient, time-varying sequences of data items produced incrementally over time, rather than being available in full before processing, in order to extract new knowledge as soon as the relevant information is collected. They are pushing traditional database, data warehousing and data mining technologies beyond their limits due to their massively increasing data volumes and demands for real-time processing, and have forced an evolution on data processing paradigms, moving from DBMSs to data stream management systems (DSMSs).

Data streams and new stream-oriented monitoring applications have the following characteristics and processing requirements [9, 25, 37, 56]:

- A data stream is an ordered, potentially unbounded sequence of data items called *stream elements*.
- Stream elements are generated by an active external source.
- The ordering is *implicit* if defined by the arrival time at the system.
- The ordering is *explicit* if stream elements provide generation time, i.e. timestamp indicating their generation by an active external source.
- Spatio-temporal data streams have explicit ordering.
- Stream elements are pushed by an active external source and arrive continuously at the system.
- DSMS has neither control over the arrival order or arrival rate of stream elements.
- Stream elements are accessed sequentially—therefore, stream element which has already arrived and which has already been processed cannot be retrieved without being explicitly stored.
- Possibility to combine real-time processing with persistent data in DBMS.
- *Stream-oriented SQL* for enabling real-time and historical analysis in a single (SQL) paradigm.
- Query over data streams runs continuously and returns new results as a new stream element arrives.

Traditional DBMSs employ a store-and-then-query data processing paradigm, i.e., data are stored in the database and *ad hoc* queries are answered in full, based on the current snapshot of the database (Fig. 1.1).

They could be used for data stream processing by loading data streams into persistent relations and repeatedly executing the same ad hoc queries over them. This approach requires that data streams need to be persisted on secondary storage devices,

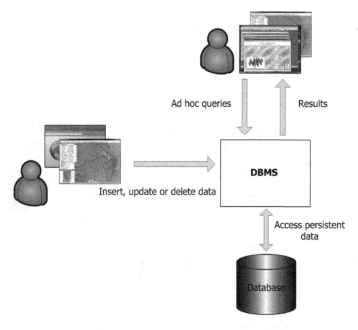

Fig. 1.1   Query processing in DBMS

i.e., disks with high latency, before they can be accessed and processed by a DBMS in main memory. The mismatch between high latency of secondary storage and low latency of main memory adds considerable delay in response time that is not acceptable to many monitoring applications. It is obvious that IFP applications do not readily fit the traditional DBMS model and its query-driven, pull-based processing paradigm, in which dynamic, transient *ad hoc* queries are typically specified, optimized and processed *once* over relatively static, persistent data [15].

Data-driven, push-based processing paradigm in DSMSs is complementary to DBMSs: the same, static, persistent queries are processed continuously over transient, dynamic, frequently changing data (Fig. 1.2).

Data stream processing model views IFP as an evolution of data processing, as supported by traditional DBMSs. Although DSMSs have their roots in DBMSs, they present significant differences: DBMSs are designed to work on persistent data where updates are relatively infrequent, while DSMSs are specialized in processing of transient data that is continuously updated. Additionally, while DBMSs run queries just once to return a complete answer, DSMSs execute the same, standing queries which run continuously and provide updated answers as new data arrives. The most DSMSs follow integrated query processing approach that runs SQL continuously and incrementally over data before that data is stored in the database. Despite differences, DSMSs resemble DBMSs, especially in the way they process incoming data through a sequence of transformations based on standard SQL operators, and all the operators defined by relational algebra [18]. Table 1.1 gives an overview of differences between DSMSs and DBMSs.

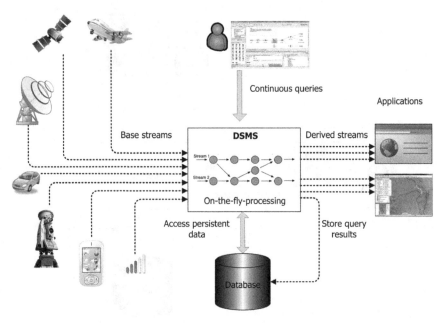

Fig. 1.2 Query processing in DSMS

Table 1.1 Differences between DSMSs and DBMSs

	DSMS	DBMS
Data	Transient streams[a]	Persistent relations
Update rates	High	Low
Data access	Sequential, one-pass	Random
Queries	Continuous	*Ad hoc*, one-time
Query results	Exact or approximate	Exact
Latency	Low	Relatively high
Processing model	Data-driven (push-based)	Query-driven (pull-based)
Query plans	Fixed	Adaptive

[a]and persistent relations

It is worth to note that aside from traditional DBMSs, three DBMS variants are related to DSMSs:

- Real-Time DBMS provides all features of traditional DBMS, while at the same time enforces applications' real-time transaction constraints. Transaction processing in real-time DBMS focuses on enforcing time constraint of transactions and ensuring temporal consistency of data [32, 38].
- In-Memory DBMS eliminates disk access by storing all data in memory and removes logical processes that are no more necessary (i.e. caching), resulting in a relatively small code footprint [30].

- Embedded DBMS is an integral part of the application or application infrastructure, and it runs with or as part application in embedded systems. Instead of providing full features of traditional DBMS, embedded DBMS provides minimal functionality such as indexing, concurrency control, logging and transactional guarantees [35, 45].

Although some of the goals for their developments were similar to those of DSMSs, they are not able to meet the requirements of IFP applications.

It is also important to note that DSMSs alone cannot entirely cover the needs for IFP: being an extension of DBMSs, they focus on producing query answers which are continuously updated to adapt to the constantly changing contents of their input data. This limitation originates from the nature of DSMSs—they are generic systems that leave the responsibility of associating a semantics to the data being processed to their clients. Complex event processing (CEP) systems are capable of processing not generic data, but *event* notifications coming from different external sources. CEP systems associate a precise semantics to the notifications of events in the external world—the CEP engine is responsible for filtering and combining event notifications to understand what is happening in terms of higher-level events. Detection and notification of complex patterns of events involving sequences and ordering relations are usually out of the scope of DSMSs. Therefore, CEP systems rely on the ability to specify composite events through event patterns that match incoming event notifications on the basis of their content and on some ordering relationships on them [18].

1.2 Data Stream Management Systems—An Overview

DSMS performs database-style query processing for IFP applications that operate in high data volume environments, and impose high throughput and low latency requirements. Due to the specific characteristics of query processing model, stream processing engines (SPEs) have a radically different architecture than that of traditional DB engine.

As we have already discussed, the queries in the continuous query (CQ) processing model execute continuously as new input data becomes available, whereas in the traditional pull-based processing model, queries are transient and input data is persistent, as illustrated in Fig. 1.2.

Traditional DBMSs apply an *outbound* processing model: data are always inserted into the DB (and indexed) before any query processing. SPEs employ an *inbound* processing model: incoming data streams directly start to flow through the query processor as they enter the DSMS [55]. If supported, read and write operations to DB are executed asynchronously. SPEs remove the storage from the critical path of processing and significantly improve performance compared to the traditional outbound processing model. Furthermore, SPEs eliminate high-overhead process context switches by executing all time-critical operations (data processing, application logic, etc.) in a single process model.

Developed SPEs share common design principles and architectural components (Fig. 1.3):

- Stream manager—handles stream I/O, including any data format conversion through built-in adaptors and informing the scheduler about arrivals of the new inputs as they are placed on the operators' queries. Stream manager also controls interaction with the network layer as external data sources typically transmit data streams over TCP or UDP sockets.
- Queue manager—ensures the availability of memory resources to support buffering inputs and outputs. It also interacts with other components to secure alternative processing techniques when memory availability guarantees cannot be upheld.
- Query executor—maintains plans for users' queries. Query plans are represented as set of operators connected with queues that buffer input data streams as well as any pending outputs.
- Query optimizer—monitors the performance of of the running query and *adaptively* determines optimal modifications to query plans and operators.
- Scheduler—determines execution strategy and an execution order to ensure efficient utilization of system resource.
- Storage manager—controls memory resources and performs asynchronous disk access and persistence.

Next, we present a list of the most influential academic prototypes and currently available commercial products.

- Aurora/Borealis—is fundamentally a data-flow system and uses the popular boxes-and-arrows paradigm found in most process flow and work flow systems. Rather

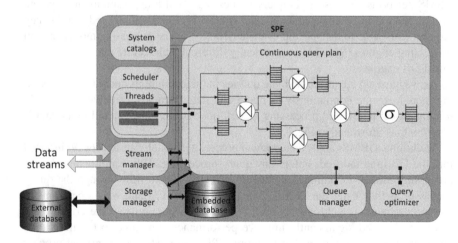

Fig. 1.3 Generic centralized architecture of an SPE [4]

than specifying continuous queries in SQL-like language, they are constructed visually in an editor, by using stream query algebra (SQuAl) operators [1]. Borealis is a distributed SPE, based on extending Aurora with distribution capabilities for a multi-site deployment [2].

- CAPE—a general-purpose DSMS that focuses on *adaptivity*, i.e. on highly reactive query operators to reduce resource usage and to increase execution efficiency, on-line query re-optimization, and adaptive operator scheduling. CAPE aims to deliver exact query answers through heterogeneous-grained adaptations with the goal to meet users QoS requirements [50].
- gStream—a geospatial data stream query engine prototype, built on the top of the spatio-temporal DBMS SECONDO [27]. gStream consists of three components: geo-stream algebra, a standalone parser, and a web-based user interface. The geo-stream algebra implements basic geospatial streaming data types and operators. The parser is tightly integrated with front-end GUI and interacts with both geo-stream and non-stream algebras [64, 65].
- InfoSphere Streams—a proprietary computing infrastructure for processing large volumes of data streams. The processing is split into basic operators called *process-ing elements*, connected to each other by streams, thus forming data-flow graphs. The system accepts rules specified using a declarative, SQL-like, SPADE language [24]. The system also allows users to write their own processing elements using a full-featured stream programming language (SPL). The modest set of spatial data types and operations are available within geospatial toolkit [10].
- MavStream—addresses stream processing holistically from a QoS perspective. MavStream SPE accepts input streams and CQs together with QoS requirements, processes CQs and monitors the output to check for QoS satisfaction [15].
- OCEANUS—a spatio-temporal DSMS prototype founded on extending TelegraphCQ with the PostGIS data types and operations along with spatio-temporal data types for supporting moving objects [21, 22].
- Oracle Stream Analytics—is a Java server including SPE-based CEP engine, which provides a declarative environment based on continuous query language (CQL) from STREAM prototype. Spatial cartridge extends CQL to support spatial data types and functions [47].
- PipelineDB—is built into the PostgreSQL core and enables data stream processing using SQL. It comes preinstalled with PostGIS and supports all other PostGIS extensions [48].
- PLACE—is a scalable location-aware database server to support CQ processing of spatio-temporal streams. The PLACE is a spatio-temporal extension of Nile query processor [29], for supporting location-aware environments [42].
- STREAM—is a general-purpose SPE with focus on adaptive approach to query processing, physical query plans, strategies against exceeding available system resources, and query visualizer [8]. The most distinctive and influential feature of STREAM project is formal specification and implemented of CQL [7].
- SAP Event Stream Processor—a proprietary event stream processor, direct descen-dant of STREAM prototype [57].

- StreamBase—a commercial real-time, stream and CEP system, based on Aurora prototype and extended with StreamSQL language [60].
- StreamInsight—is a proprietary platform for developing and deploying streaming applications. The query language supported by streaming engine is Language Integrated Query (LINQ) which supports a number of conventional SQL operators. Extensibility framework of StreamInsight enables a seamless integration of Spatial Library into query pipeline to support the online processing of spatio-temporal data streams [41].
- Stream Mill—is a general purpose DSMS prototype, focused on lifting certain limitations imposed by SQL-based languages. Expressive Stream Language (ESL) extends SQL into a Turing-complete language by means of user defined aggregates (UDA), and supports slides and windows on arbitrary aggregates, as opposed to only built-in ones [58].
- TelegraphCQ—an open-source prototype based on the PostgreSQL database system. Telegraph consists of an extensible set of composable data-flow modules or operators that can be composed into multi-step data-flows, and supports communication via either asynchronous *push* or synchronous *pull* modalities. Data-flows are initiated either via an ad hoc query language StreaQuel, or via a scripting language for representing data-flow graphs explicitly [16].

This is by no means an exhaustive list, but it gives some idea of their capabilities as well as limitations for spatio-temporal stream processing.

1.3   Data Stream Mining and Knowledge Discovery—An Overview

The explosive growth in streaming data has generated an urgent need for new techniques and tools for transforming the massive amounts of transient, dynamic data into knowledge. Knowledge discovery is the process of identifying valid, interesting, previously unknown understandable patterns representing knowledge implicitly stored in massive data repositories (DB, DW, Web) or data streams. Data mining is the mathematical core of knowledge discovery process, involving the deductive algorithms that explore data, develop mathematical models and discover significant patters. Traditional database query languages, statistical methods and CQL are not capable to discover interesting patterns in very large data sets. Therefore, both KDD and KDDS encompass principle and techniques from statistics, machine learning, pattern recognition, artificial intelligence, knowledge-based systems and data visualization [31].

Data stream mining, the crucial step of the knowledge discovery process, is concerned with discovering and extracting knowledge structures and patterns in data streams. Research on data mining and knowledge discovery from data streams is relatively new and primarily focused on traditional, non-location-aware data streams. While knowledge discovery from traditional data streams (KDDS) is a hot research topic, research in spatial and spatio-temporal domains has only just started. The evolution and rising of spatial and spatio-temporal KDDS is analogous to the evolution of spatial and spatio-temporal DBMS described at the beginning, and in fact, many of the critical issues related to spatio-temporal KDDS are still open research challenges.

However, extracting interesting and useful patterns from spatial data sets is more difficult than extracting corresponding patterns from traditional numeric and categorical data due to the complexity of spatial data types, spatial relationships, spatial auto-correlation, and non-linearity. Both spatial and spatio-temporal data require consideration of spatial auto-correlation, and spatial relationships and constraints in the model building. Spatial auto-regression (SAR) and Markov random field (MRF) are two major approaches for incorporating spatial dependence (or neighborhood relationships) into the classification and prediction problems: [53, 54, 62].

The complexity of spatial data and inherent spatio-temporal relationships limits the usefulness of traditional data mining techniques for extracting spatio-temporal patterns. As an motivation example, let us consider the task of mining spatio-temporal patterns of four objects freely moving in 2D Euclidean space over the last 20 time units (Fig. 1.4).

As objects move through the space, they generate trajectory data streams. Obviously, neither conventional or spatial data mining methods *per se* are not capable to discover spatio-temporal patterns like *flock*, *meet*, *periodic*, *frequent location*, etc. Moreover, due to well-known limitations imposed by data stream processing such as limited memory resources, high speed data arrival, nearly real-time processing, and one-pass techniques, even available and well established multi-step methodologies

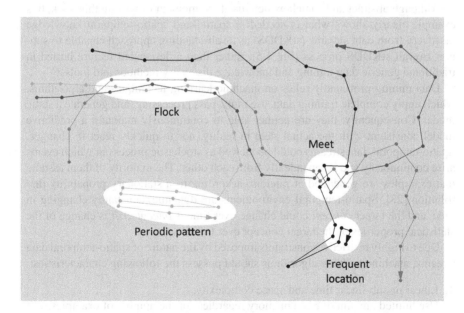

Fig. 1.4 An example of spatio-temporal patterns [26]

and multi-scan algorithms, suitable for KDD, cannot be smoothly applied to data streams. In fact, it would be ideal to have a *continuous DMQL* at our disposal, similar to that of [49, 61] and inspired by [34]:

//create data mining task
create data mining *example;*

//create data
create data *example.stStream* **as**
(select * **from** *streamRowData* ***[range 20 seconds]*)**;**

//create model by choosing algorithm
create model *example.clusters* **using cluTraStream as**
(select *s.id, s.point* **from** *stStream s*
where cluTraStream.minLineSegments = 3 **and cluTraStream.epsilon** = 10);

//create relation between data and model
create entailment stream *example.trajectoriesClustering* **as**
(select *s.id, s.point, c.id, c.point* **from** *example.stStream s, example.clusters c*);

//finally, create flocks
create model *example.flocks* **using flockPattern as**
(select *s.id, s.point* **from** *example.stStream s, example.trajectoryClustering t*);

Although abstract and somehow fictious at the moment of writing this book, this example clearly shows what is needed: a spatial and spatio-temporal knowledge discovery from data streams (stKDDS) as a self-standing approach capable to support overall stKDDS process (Fig. 1.5), rather than a descended feature buried in traditional generic data mining and knowledge discovery methods and tools.

Data mining profoundly relies on machine learning techniques and algorithms, which imply complete training data, use multi-pass processing, and generate a static model. Consequently, they are neither able to continuously maintain a predictive model consistent with the actual state in reality, nor to quickly react to changes. Spatio-temporal data streams could be viewed as stochastic process in which events arise continuously and independently from each other. The majority of them assume that examples are generated at random, according to a stationary probability distribution [23]. Spatio-temporal environments are dynamic, constantly changing in time, and the target *concept* could change over time. *Concept drift* is change of the statistical properties of the target concept over time.

Due to highly restrictive constraints imposed by the nature of spatio-temporal data streams, machine learning algorithms should possess the following characteristics:

1. Linear or sub-linear time and space complexity.
2. Use limited amount of main memory regardless of the number of examples.
3. Create a model using a single scan over the training data.

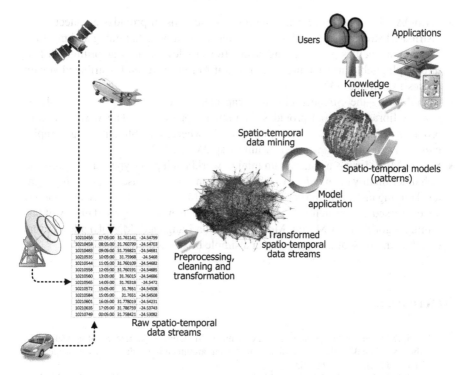

Fig. 1.5 The process of spatio-temporal knowledge discovery in data streams

4. Capability to deal with concept drift.
5. Iterative updates to the model.

   While knowledge discovery from data streams (KDDS) is hot research topic, complementary research in spatial and spatio-temporal domain is in its infancy, both in computer science and geospatial engineering. Indeed, most of the critical issues related to stKDDS are still open as current research challenges.

   We conclude this chapter by presenting a list of relevant data stream mining prototypes and currently available commercial products.

- InfoSphere Streams—includes mining toolkit which consists of a set of machine learning algorithms for association rules, classification, regression and clustering. Data stream mining algorithms use predefined training model created from persistent data and stored in PMML format [10].
- MAIDS—a prototype consisting of five functional modules, including stream data classifier, stream pattern finder and stream cluster analyser [14].
- MOA—an open source workbench for data stream mining written in Java, founded on a collection of machine learning algorithms: classification, regression, and clustering [11].

- SAMOA—is a platform for mining big data streams. It provides a collection of distributed streaming algorithms for the most common data mining and machine learning, as well as programming abstractions to develop new algorithms. It features a pluggable architecture that allows it to run on several distributed stream processing engines [43].
- RapidMiner—the Streams Plugin encapsulates the streaming runtime of the *streams* library [12] and provides data stream processing. The *streams* library provides a simple execution runtime by itself, whereas the Streams Plugin implements an execution environment within RapidMiner [13].
- Stream Mill Miner—a data stream mining workbench prototype that supports the rich functionality of an *inductive* DSMS and combines the ease of specifying high-level mining tasks with the performance of a DSMS. Knowledge discovery queries are expressed as user-defined aggregates (UDAs) using an integrated library of fast mining algorithms. A Mining Model Definition Language (MMDL) allows users to define the flow of mining tasks, in a simple boxes-and-arrows GUI [59].

References

1. Abadi, D.J., Carney, D., Çetintemel, U., Cherniack, M., Convey, C., Lee, S., Stonebraker, M., Tatbul, N., Zdonik, S.B.: Aurora: a new model and architecture for data stream management. VLDB J. **12**(2), 120–139 (2003)
2. Abadi, D.J., Ahmad, Y., Balazinska, M., Çetintemel, U., Cherniack, M., Hwang, J.H., Lindner, W., Maskey, A., Rasin, A., Ryvkina, E., Tatbul, N., Xing, Y., Zdonik, S.B.: The design of the Borealis stream processing engine. In: CIDR. pp. 277–289 (2005)
3. Aberer, K., Franklin, M.J., Nishio, S. (eds.): In: Proceedings of the 21st International Conference on Data Engineering, ICDE 2005, 5–8 April 2005, Tokyo, Japan. IEEE Computer Society (2005)
4. Ahmad, Y., Çetintemel, U.: Data stream management architectures and prototypes. In: Liu and Özsu [39], pp. 639–643
5. Aitchison, A.: Pro Spatial with SQL Server 2012. Apress Media LLC, New York (2012)
6. de Almeida, V.T., Güting, R.H., Behr, T.: Querying moving objects in SECONDO. In: Mobile Data Management. pp. 47–51. IEEE Computer Society (2006)
7. Arasu, A., Babu, S., Widom, J.: The CQL continuous query language: semantic foundations and query execution. Int J Large Databases **15**(2), 121–142 (2006)
8. Arasu, A., Babcock, B., Babu, S., Cieslewicz, J., Datar, M., Ito, K., Motwani, R., Srivastava, U., Widom, J.: STREAM: The Stanford data stream management system. Technical Report 2004-20, Stanford InfoLab (2004). http://ilpubs.stanford.edu:8090/641/
9. Babcock, B., Babu, S., Datar, M., Motwani, R., Widom, J.: Models and issues in data stream systems. In: Popa, L., Abiteboul, S., Kolaitis, P.G. (eds.) PODS. pp. 1–16. ACM (2002)
10. Ballard, C., Brandt, O., Devaraju, B., Farrell, D., Foster, K., Howard, C., Nicholls, P., Pasricha, A., Rea, R., Schulz, N., Shimada, T., Thorson, J., Tucker, S., Uleman, R.: IBM InfoSphere Streams: Accelerating Deployments with Analytic Accelerators. IBM (2014)
11. Bifet, A., Holmes, G., Pfahringer, B., Kranen, P., Kremer, H., Jansen, T., Seidl, T.: MOA: massive online analysis, a framework for stream classification and clustering. J. Mach. Learn. Res. Proc. Track **11**, 44–50 (2010)
12. Bockermann, C.: The stream framework. http://www-ai.cs.uni-dortmund.de/SOFTWARE/streams/index.html (2015)

13. Bockermann, C., Blom, H.: Processing data streams with the RapidMiner streams plugin. http://www.jwall.org/streams/doc/rapidminer.html (2015)
14. Cai, Y.D., Clutter, D., Pape, G., Han, J., Welge, M., Auvil, L.: MAIDS: Mining alarming incidents from data streams. In: Weikum, G., König, A.C., Deßloch, S. (eds.) SIGMOD Conference. pp. 919–920. ACM (2004)
15. Chakravarthy, S., Jiang, Q.: Stream Data Processing: A Quality of Service Perspective - Modeling, Scheduling, Load Shedding, and Complex Event Processing, Advances in Database Systems, vol. 36. Kluwer (2009)
16. Chandrasekaran, S., Cooper, O., Deshpande, A., Franklin, M.J., Hellerstein, J.M., Hong, W., Krishnamurthy, S., Madden, S., Raman, V., Reiss, F., Shah, M.A.: TelegraphCQ: Continuous dataflow processing for an uncertain world. In: CIDR (2003)
17. Chen, C.X.: Spatio-temporal query languages. In: Shekhar and Xiong [54], pp. 1125–1128
18. Cugola, G., Margara, A.: Processing flows of information: From data stream to complex event processing. ACM Comput. Surv. 44(3), 15:1–15:60 (2012)
19. Frentzos, E., Pelekis, N., Ntoutsi, I., Theodoridis, Y.: Mobility, Data Mining and Privacy - Geographic Knowledge Discovery, chap. Trajectory Database Systems, pp. 151–187. Springer, Berlin (2008)
20. Galić, Z.: Geospatial Databases. Golden Marketing-Tehnička knjiga, Zagreb (2006). [in Croatian]
21. Galić, Z., Mešković, E., Križanović, K., Baranović, M.: Oceanus: a spatio-temporal data stream system prototype. In: Proceedings of the Third ACM SIGSPATIAL International Workshop on GeoStreaming. pp. 109–115. IWGS '12, ACM, New York, NY, USA (2012). http://doi.acm.org/10.1145/2442968.2442982
22. Galić, Z., Baranović, M., Križanović, K., Mešković, E.: Geospatial data streams: Formal framework and implementation. Data & Knowledge Engineering 91, 1–16 (2014). http://dx.doi.org/10.1016/j.datak.2014.02.002
23. Gama, J.: Knowledge Discovery from Data Streams, 1st edn. Chapman & Hall/CRC, Boca Raton, FL, USA (2010)
24. Gedik, B., Andrade, H., Wu, K.L., Yu, P.S., Doo, M.: Spade: the system s declarative stream processing engine. In: Wang, J.T.L. (ed.) SIGMOD Conference. pp. 1123–1134. ACM (2008)
25. Golab, L., Özsu, M.T.: Data Stream Management.Synthesis Lectures on Data Management. Morgan Claypool Publishers, San Rafael, CA (2010)
26. Gudmundsson, J., Laube, P., Wolle, T.: Computational movement analysis. In: Springer Handbook of Geographic Information, pp. 725–741. Springer-Verlag, Berlin Heidelberg (2012)
27. Güting, R.H., de Almeida, V.T., Ansorge, D., Behr, T., Ding, Z., Höse, T., Hoffmann, F., Spiekermann, M., Telle, U.: SECONDO: An extensible DBMS platform for research prototyping and teaching. In: Aberer et al. [3], pp. 1115–1116
28. Güting, R.H., Schneider, M.: Moving Objects Databases. Morgan Kaufmann, San Francisco, CA (2005)
29. Hammad, M.A., Mokbel, M.F., Ali, M.H., Aref, W.G., Catlin, A.C., Elmagarmid, A.K., Eltabakh, M.Y., Elfeky, M.G., Ghanem, T.M., Gwadera, R., Ilyas, I.F., Marzouk, M.S., Xiong, X.: Nile: A query processing engine for data streams. In: Özsoyoglu, Z.M., Zdonik, S.B. (eds.) ICDE. p. 851. IEEE Computer Society (2004)
30. Han, H., il Jin, S.: A main memory based spatial DBMS: Kairos. In: Lee, S.G., Peng, Z., Zhou, X., Moon, Y.S., Unland, R., Yoo, J. (eds.) DASFAA (2). Lecture Notes in Computer Science, vol. 7239, pp. 234–242. Springer (2012)
31. Han, J., Kamber, M., Pei, J.: Data Mining: Concepts and Techniques, 3rd edn. Morgan Kaufmann Publishers Inc., San Francisco, CA, USA (2011)
32. Hansson, J., Xiong, M.: Real-time transaction processing. In: Liu and Özsu [39], pp. 2344–2348
33. InfoLab: HERMES. http://hermes-mod.java.net (2015)
34. Johnson, T., Lakshmanan, L.V.S., Ng, R.T.: The 3w model and algebra for unified data mining. In: El Abbadi, A., Brodie, M.L., Chakravarthy, S., Dayal, U., Kamel, N., Schlageter, G., Whang, K.Y. (eds.) VLDB. pp. 21–32. Morgan Kaufmann (2000)

35. Kang, W., Son, S.H., Stankovic, J.A.: Design, implementation, and evaluation of a QoS-aware real-time embedded database. IEEE Trans. Comput. **61**(1), 45–59 (2012)
36. Koubarakis, M., Sellis, T.K., Frank, A.U., Grumbach, S., Güting, R.H., Jensen, C.S., Lorentzos, N.A., Manolopoulos, Y., Nardelli, E., Pernici, B., Schek, H.J., Scholl, M., Theodoulidis, B., Tryfona, N. (eds.): Spatio-Temporal Databases: The CHOROCHRONOS Approach, Lecture Notes in Computer Science, vol. 2520. Springer (2003)
37. Krämer, J., Seeger, B.: Semantics and implementation of continuous sliding window queries over data streams. ACM Trans. Database Syst. 34(1) (2009)
38. Lindström, J.: Real time database systems. In: Wah, B.W. (ed.) Wiley Encyclopedia of Computer Science and Engineering. Wiley, New York (2008)
39. Liu, L., Özsu, M.T. (eds.): Encyclopedia of Database Systems. Springer, US (2009)
40. Meng, X., Chen, J.: Moving Objects Management: Models. Techniques and Applications. Tsinghua University Press and Springer-Verlag, Beijing and Berlin Heidelberg (2010)
41. Miller, J., Raymond, M., Archer, J., Adem, S., Hansel, L., Konda, S., Luti, M., Zhao, Y., Teredesai, A., Ali, M.H.: An extensibility approach for spatio-temporal stream processing using Microsoft StreamInsight. In: Pfoser, D., Tao, Y., Mouratidis, K., Nascimento, M.A., Mokbel, M.F., Shekhar, S., Huang, Y. (eds.) SSTD. Lecture Notes in Computer Science, vol. 6849, pp. 496–501. Springer (2011)
42. Mokbel, M.F., Xiong, X., Hammad, M.A., Aref, W.G.: Continuous query processing of spatio-temporal data streams in PLACE. GeoInformatica **9**(4), 343–365 (2005)
43. Morales, G.D.F., Bifet, A.: SAMOA: scalable advanced massive online analysis. J. Mach. Learn. Res. 16, 149–153 (2015). http://dl.acm.org/citation.cfm?id=2789277
44. Murray, C.: Oracle Spatial and Graph Developer's Guide. Oracle (2014)
45. Nori, A.: Mobile and embedded databases. IEEE Data Eng. Bull. **30**(3), 3–12 (2007)
46. Obe, R., Hsu, L., Ramsey, P.: PostGIS in Action. Manning Publications, Greenwich, CT (2012)
47. Oracle: Oracle Stream Analytics. http://www.oracle.com/technetwork/middleware/complex-event-processing (2016)
48. PipelineDB: PipelineDB. www.pipelinedb.com (2015)
49. Renso, C., Trasarti, R.: Understanding human mobility using mobility data mining. Mobility Data-Modeling. Management, and Understanding, pp. 127–147. Cambridge University Press, New York (2013)
50. Rundensteiner, E.A., Ding, L., Sutherland, T.M., Zhu, Y., Pielech, B., Mehta, N.: CAPE: Continuous query engine with heterogeneous-grained adaptivity. In: Nascimento, M.A., Özsu, M.T., Kossmann, D., Miller, R.J., Blakeley, J.A., Schiefer, K.B. (eds.) VLDB. pp. 1353–1356. Morgan Kaufmann (2004)
51. Shekhar, S., Chawla, S.: Spatial Databases-A Tour. Prentice Hall, Upper Saddle River, NJ (2003)
52. Shekhar, S., Xiong, H. (eds.): Encyclopedia of GIS. Springer, New York (2008)
53. Shekhar, S., Vatsavai, R.R., Celik, M.: Spatial and spatiotemporal data mining: Recent advances. In: Next Generation of Data Mining, (1st edn.) pp. 549–584. Chapman & Hall/CRC (2008)
54. Shekhar, S., Evans, M.R., Kang, J.M., Mohan, P.: Identifying patterns in spatial information: a survey of methods. Wiley Interdisc. Rew. Data. Min. Knowl. Discov. **1**(3), 193–214 (2011)
55. Stonebraker, M., Çetintemel, U.: "One size fits all": an idea whose time has come and gone. In: Aberer et al. [3], pp. 2–11
56. Stonebraker, M., Çetintemel, U., Zdonik, S.B.: The 8 requirements of real-time stream processing. SIGMOD Record **34**(4), 42–47 (2005)
57. Sybase: SAP Event Stream Processor. http://go.sap.com/product/data-mgmt/complex-event-processing.html (2016)
58. Thakkar, H., Zaniolo, C.: Introducing Stream Mill: User-Guide to the Data Stream Management System, its Expressive Stream Language ESL, and the Data Stream Mining Workbench SMM. Computer Science Department, UCLA (October (2010)
59. Thakkar, H., Laptev, N., Mousavi, H., Mozafari, B., Russo, V., Zaniolo, C.: SMM: A data stream management system for knowledge discovery. In: Abiteboul, S., Böhm, K., Koch, C., Tan, K.L. (eds.) ICDE. pp. 757–768. IEEE Computer Society (2011)

60. TIBCO Software Inc.: TIBCO StreamBase. http://www.streambase.com (2016)
61. Trasarti, R., Giannotti, F., Nanni, M., Pedreschi, D., Renso, C.: A query language for mobility data mining. IJDWM **7**(1), 24–45 (2011)
62. Vatsavai, R.R., Shekhar, S., Burk, T.E., Bhaduri, B.L.: *Miner: a spatial and spatiotemporal data mining system. In: Aref, W.G., Mokbel, M.F., Schneider, M. (eds.) GIS. p. 86. ACM (2008)
63. Xiong, X., Mokbel, M.F., Aref, W.G.: Spatio-temporal database. In: Shekhar and Xiong [54], pp. 1114–1115
64. Zhang, C.: gStream. http://powerranger.cse.unt.edu/gstream (2013)
65. Zhang, C., Huang, Y., Griffin, T.: Querying geospatial data streams in SECONDO. In: Agrawal, D., Aref, W.G., Lu, C.T., Mokbel, M.F., Scheuermann, P., Shahabi, C., Wolfson, O. (eds.) GIS. pp. 544–545. ACM (2009)

Chapter 2
Spatio-Temporal Continuous Queries

Abstract Spatio-temporal stream processing in general refers to a class of software systems for processing of high volume spatio-temporal data streams with very low latency, i.e. in near real-time. Motivated by the limitation of DBMS, the database community developed data stream management systems (DSMSs), as a new class of management systems oriented toward processing large data streams in a near real-time. Despite differences these between these two classes of management systems, DSMSs resemble DBMSs—they process data streams using SQL and operators defined by the relational algebra. This chapter gives an insight into spatio-temporal stream processing at conceptual level, i.e. from the DSMS user perspective.

Keywords Data streaming · Data stream architectures · Spatio-temporal data streams · GeoStreaming · Continuous query processing · Stream windows

2.1 Foundation of Continuous Query Processing

When processing data streams, there are two inherent temporal domains to consider:

- *Event* time,[1] which is the time at which the event itself actually occurred in the real world.
- *Processing* time, which is the current time according to the system clock, at which an event is observed during processing.

Most DSMS are founded on extensions of the relational model and corresponding query languages. Consequently, data stream items could be viewed as relational tuples with one important distinction: they are *time ordered*. The ordering is defined either explicitly by event time (a.k.a valid time) or implicitly by processing time. Similarly to [9], we define *time domain*, *time instant* and *time interval* as follows:

Definition 2.1 (*Time Domain*) A time domain $\mathbb{T}$ is a pair $(T; \leqslant)$ where T is non-empty set of discrete time instants and $\leqslant$ is total order on T.

Definition 2.2 (*Time Instant*) A time instant τ is any value from T, i.e. $\tau \in T$.

---

[1] The terms *event time* and *valid time* are often used interchangeably.

© The Author(s) 2016
Z. Galić, *Spatio-Temporal Data Streams*, SpringerBriefs
in Computer Science, DOI 10.1007/978-1-4939-6575-5_2

Definition 2.3 (*Time Period*) A time period represents extend in time defined by the temporal positions of the time instants at which it begins (τ_{begin}) and ends (τ_{end}).

Discrete time domain implies that every time instant has an immediate successor (except the last, if any) and immediate predecessor (except the first, if any).

Definition 2.4 (*Time Interval*) A time interval consists of all distinct time instants $\tau \in T$ and could be open, closed, left-closed or right-closed:

$$(\tau_{begin}, \tau_{end}) = \{\tau \in \mathbb{T} \mid \tau_{begin} < \tau < \tau_{end}\},$$
$$[\tau_{begin}, \tau_{end}] = \{\tau \in \mathbb{T} \mid \tau_{begin} \leq \tau \leq \tau_{end}\},$$
$$[\tau_{begin}, \tau_{end}) = \{\tau \in \mathbb{T} \mid \tau_{begin} \leq \tau < \tau_{end}\},$$
$$(\tau_{begin}, \tau_{end}] = \{\tau \in \mathbb{T} \mid \tau_{begin} < \tau \leq \tau_{end}\}.$$

Spatio-temporal data streams have two distinct features which differentiate them from conventional data streams based on relational model:

- Event time of data stream tuple is defined by temporal attribute $\mathscr{A}_{\theta}$.
- Shape and location of an object of interest described by a data stream tuple defined by spatial attribute $\mathscr{A}_{\sigma}$.

First feature implies that each data stream tuple has *event timestamp* [27] generated by the source, which classifies spatio-temporal streams into a class of *explicitly timestamped data streams* [31].

Spatial domain is a set of homogeneous object structures (values) which provide a fundamental abstraction for modelling the geometric structure of real-world phenomena in space. Points, lines, polygons and surfaces are the most popular and fundamental abstractions for mapping a geometric structure from 3D space into 2D space [44]. In order to locate object in space, the embedding space must be defined as well. The formal treatment of spatial domain requires a definition of mathematical space. Although Euclidean $\mathscr{R}^2$ embedding space seems to be dominant, in some cases other spaces (metric, vector, topological) are more important.

Simple object structures (point, continuous line, simple polygon) are not closed under the geometric set operations (difference, intersection, union). This means that geometric set operations can produce *complex* spatial objects (multipoints, multilines, multipolygons, polygons with holes, etc.). For this reason, spatial domain should include spatial objects with complex structure.

Definition 2.5 (*Spatial Domain*) A spatial domain $\mathscr{D}_{\sigma}$ is a set of spatial objects with simple or complex structure.

At first glance, complex spatial objects may appear to be overwhelming in DSMS context. However, we want to relay on and explore extensive research on abstract spatial data types in spatial DBMS and GIS. More specifically, we are going to follow *abstract spatial data types* framework, as defined in [24, 25], where the internal structure of a spatial object is hidden from the user, and could only be accessed through a set of predefined operations (Fig. 2.1). Spatial domain, even though consisting of simple object structures only, is obviously not atomic but rather structured,

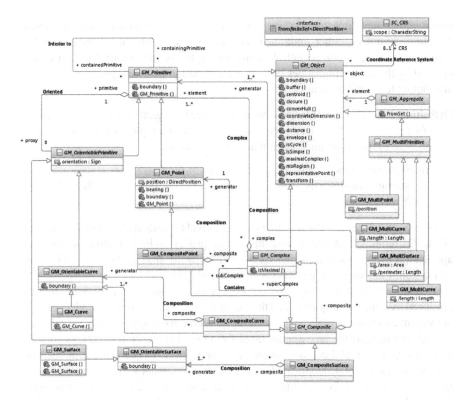

Fig. 2.1 ISO 19107 abstract spatial data types [24]

and as a consequence, spatio-temporal data streams rely on object-relational (or object-oriented) paradigm.

Having defined the time and spatial domains, we define spatio-temporal data stream schema:

Definition 2.6 (*Spatio-Temporal Data Stream Schema*) A spatio-temporal data stream schema $\Sigma_{\sigma\theta}$ is represented as a set of attributes $\langle \mathcal{A}_1, \mathcal{A}_2, \ldots, \mathcal{A}_n \rangle$ of finite arity n. One of the attributes (denoted by $\mathcal{A}_\sigma$) has associated spatial domain $\mathcal{D}_\sigma$, and one of the attributes (denoted by $\mathcal{A}_\theta$) has associated temporal domain $\mathcal{D}_\theta$, i.e. $\mathbb{T}$. The values of other $n-2$ attributes are drawn either from atomic type domain $\mathcal{D}_{\alpha_i}$ or complex type domain $\mathcal{D}_{\chi_i}$.

Finally, spatio-temporal data stream is defined in the following way:

Definition 2.7 (*Spatio-Temporal Data Stream*) A spatio-temporal data stream $\mathcal{S}_{\sigma\theta}$ is a possibly infinite sequence of tuples belonging to the schema of $\Sigma_{\sigma\theta}$ and ordered by the increasing values of $\mathcal{A}_\theta$.

A spatio-temporal stream tuple t represents an *event*, i.e. an instantaneous fact capturing information that occurred in real-world at time instant τ, defined by event timestamp. Event timestamp offers a unique time indication for each tuple, and therefore cannot be either undefined (i.e., event timestamp cannot have a `null` value) or mutable. In the sequel, we consider explicitly timestamped data streams, ordered by the increasing values of their event timestamps. Analogously to [42], we also define temporal ordering:

Definition 2.8 (*Temporal Order*) A temporal order is surjective (many-to-one) mapping $f_\Omega : \mathscr{D}_{\mathscr{S}_{\sigma\theta}} \twoheadrightarrow \mathbb{T}$ from data type domain $\mathscr{D}_{\mathscr{S}_{\sigma\theta}}$ of the tuples belonging to a data stream $\mathscr{S}_{\sigma\theta}$ to time domain $\mathbb{T}$, such that following holds:

1. Timestamp existence: $\forall s \in \mathscr{S}_{\sigma\theta},\ \exists \tau \in \mathbb{T},\ such\ that f_\Omega(s) = \tau$.
2. Timestamp monotonicity: $\forall s_1, s_2 \in \mathscr{S}_{\sigma\theta},\ if s_1.\mathscr{A}_\theta \le s_2.\mathscr{A}_\theta,\ then f_\Omega(s_1) \le f_\Omega(s_2)$.

DSMS could tag each stream tuple with its arrival timestamp, using system's local clock. A stream with system timestamps can be processed like a regular stream with application event timestamps, but we should be aware that application event time and system time are not necessarily synchronized [29].

We distinguish between *raw* streams produced by the sources and *derived* streams produced by continuous queries and their operators. In either case, we model individual stream elements as object-relational tuples with a fixed spatio-temporal schema.

The raw streams are an essential input for a broad range of applications such as traffic management and control, routing, and navigation. To become useful, the raw streams have to be related to the underlying transportation network by means of *map-matching* algorithms. For example, the map-matching is one of the key operations found in Intelligent Transportation Systems—a map-matching UDF could be applied on raw stream to produce derived stream.

2.1.1 Running Example

As a running example,[2] let us consider the raw stream generated by GPS and speed sensors embedded into a moving object:

```
CREATE STREAM gpsStream (
    id          VARCHAR(8),
    lat         REAL,       // Latitude
    lon         REAL,       // Longitude
    elevation   SMALLINT,   // Ellipsoidal height
    speed       REAL,       // Speed [km/h]
    timestamp   TIMESTAMP VALIDTIME
)
```

[2]The notation used in this book is close to the notation used in [15, 16], which is itself based on TelegraphCQ [14], PostGIS [43] and SQL/MM—Spatial [25]. Due to the simplicity and clarity of the syntax, as well as to avoid possible confusion with spatio-temporal data types, prefix `ST_` has been omitted.

```
ORDERED BY timestamp;
ALTER STREAM gpsStream ADD WRAPPER gpsWrapper;
```

Stream might have multiple attributes of TIMESTAMP type, but only one should have VALIDTIME constraint. This constraint implicitly determines the read-only attribute according to which the stream is ordered. Wrappers are user-defined data acquisition functions that transform the sequence of bytes into a raw stream, and ALTER STREAM associates the raw stream with a wrapper.

An example of a sequence of gpsStream tuples (Fig. 2.2):

```
. . .
"W-45084A" 48.20781333 16.43832500 221 73.2 2015-10-19 11:50:30
"W-45084A" 48.20795500 16.43853167 221 79.2 2015-10-19 11:50:31
"W-45084A" 48.20809167 16.43873667 220 77.4 2015-10-19 11:50:32
"W-45084A" 48.20823500 16.43894667 220 80.3 2015-10-19 11:50:33
"W-45084A" 48.20838167 16.43916333 220 82.5 2015-10-19 11:50:34
"W-45084A" 48.20851667 16.43936000 220 75.4 2015-10-19 11:50:35
"W-45084A" 48.20865833 16.43957167 220 80.1 2015-10-19 11:50:36
"W-45084A" 48.20881000 16.43978667 219 83.6 2015-10-19 11:50:37
"W-45084A" 48.20894667 16.43997667 219 74.7 2015-10-19 11:50:38
"W-45084A" 48.20908667 16.44017833 219 77.8 2015-10-19 11:50:39
"W-45084A" 48.20923167 16.44038333 218 79.8 2015-10-19 11:50:40
"W-45084A" 48.20937667 16.44059167 218 80.5 2015-10-19 11:50:41
. . .
```

Derived stream with position modelled as a point on WGS84 ellipsoid, can be quite natural for applications involving spatio-temporal objects whose moving is related to the Earth's surface: airplanes, tankers, combat aircrafts, cruise missiles, drones, etc.:

```
CREATE STREAM movingObjectWGS84 AS
SELECT id,
       SetSRID(Point(lon,lat,elevation),4326)::GEOGRAPHY
       AS wgsPosition,
       speed,
       timestamp
FROM gpsStream;
```

Parameter 4326 is EPSG[3] identifier of WGS 84 spatial reference system, and : : is shorthand for type casting.

Here is a derived sequence of movingObjectWGS84 tuples:

```
. . .
"W-45084A" POINT(48.20781333 16.43832500 221) 73.2 2015-10-19 11:50:30
"W-45084A" POINT(48.20795500 16.43853167 221) 79.2 2015-10-19 11:50:31
"W-45084A" POINT(48.20809167 16.43873667 220) 77.4 2015-10-19 11:50:32
"W-45084A" POINT(48.20823500 16.43894667 220) 80.3 2015-10-19 11:50:33
"W-45084A" POINT(48.20838167 16.43916333 220) 82.5 2015-10-19 11:50:34
"W-45084A" POINT(48.20851667 16.43936000 220) 75.4 2015-10-19 11:50:35
```

---

[3] http://spatialreference.org/ref/epsg/4326/.

```
"W-45084A" POINT(48.20865833 16.43957167 220) 80.1 2015-10-19 11:50:36
"W-45084A" POINT(48.20881000 16.43978667 219) 83.6 2015-10-19 11:50:37
"W-45084A" POINT(48.20894667 16.43997667 219) 74.7 2015-10-19 11:50:38
"W-45084A" POINT(48.20908667 16.44017833 219) 77.8 2015-10-19 11:50:39
"W-45084A" POINT(48.20923167 16.44038333 218) 79.8 2015-10-19 11:50:40
"W-45084A" POINT(48.20937667 16.44059167 218) 80.5 2015-10-19 11:50:41
. . .
```

We may define another derived data stream with a position modelled as a point in two-dimensional Euclidean space as:

```
CREATE STREAM movingObject AS
SELECT id,
       Force_2D(Transform(wgsPosition::GEOMETRY,3416))
       AS position,
       speed,
       timestamp
FROM movingObjectWGS84;
```

Function `Transform` transforms a point on WGS84 ellipsoid (`wgsPoint`) to the specified spatial reference system. Parameter 3416 is a unique identifier (SRID) used to unambiguously identify EPSG:3416[4] spatial reference system, which incorporates European Terrestrial Reference System 1989 (ETRS89) and Lambert projection.

Finally, here is a tuple sequence of derived `movingObject` stream:

```
. . .
"W-45084A" POINT(630656.02 483284.08) 73.2 2015-10-19 11:50:30
"W-45084A" POINT(630670.74 483300.43) 79.2 2015-10-19 11:50:31
"W-45084A" POINT(630685.35 483316.22) 77.4 2015-10-19 11:50:32
"W-45084A" POINT(630700.30 483332.76) 80.3 2015-10-19 11:50:33
"W-45084A" POINT(630715.74 483349.70) 82.5 2015-10-19 11:50:34
"W-45084A" POINT(630729.74 483365.28) 75.4 2015-10-19 11:50:35
"W-45084A" POINT(630744.82 483381.64) 80.1 2015-10-19 11:50:36
"W-45084A" POINT(630760.17 483399.13) 83.6 2015-10-19 11:50:37
"W-45084A" POINT(630773.62 483414.87) 74.7 2015-10-19 11:50:38
"W-45084A" POINT(630787.97 483431.02) 77.8 2015-10-19 11:50:39
"W-45084A" POINT(630802.54 483447.74) 79.8 2015-10-19 11:50:40
"W-45084A" POINT(630817.36 483464.46) 80.5 2015-10-19 11:50:41
. . .
```

Previous two examples illustrate the concept of derived streams. Of course, it would be possible to define corresponding spatio-temporal raw streams in straight way as:

---

[4]http://spatialreference.org/ref/epsg/3416/.

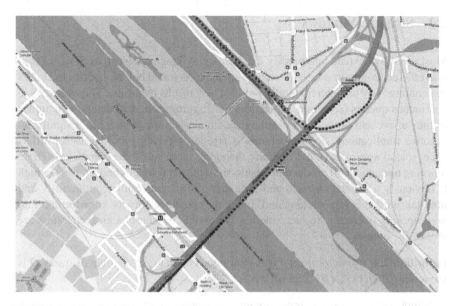

Fig. 2.2 Visualisation of a spatio-temporal stream

```
CREATE STREAM movingObjectWGS84 (
    id          VARCHAR(8),
    wgsPosition GEOGRAPHY(POINT,4326)
    speed       REAL,
    timestamp   TIMESTAMP VALIDTIME
)
ORDERED BY timestamp;
ALTER STREAM movingObjectWGS84 ADD WRAPPER moWrapperWGS84;
```

and

```
CREATE STREAM movingObject (
    id        VARCHAR(8),
    position  GEOMETRY(POINT, 3416)
    speed     REAL,
    timestamp TIMESTAMP VALIDTIME
)
ORDERED BY timestamp;
ALTER STREAM movingObject ADD WRAPPER moWrapper;
```

In both cases, complete logic of data acquisition and transformation of bytes into a row streams is encapsulated into corresponding wrappers.

2.2   Stream Windows

Most of DSMS extend and modify database query language (such as SQL) to support efficient continuous queries on data streams. As stated before, queries in DSMS run continuously and incrementally produce new results over time. The operators in continuous queries (selection, projection, join, aggregation, etc.) compute on tuples as they arrive and do not suppose the data stream is finite, which has a significant negative implications. Some operators (Cartesian product, join, union, set difference, spatial aggregation, etc.) require the entire input sets to be completed. These *blocking* operators will produce no results until the data streams end (if ever), which is obviously a serious obstacle and limitation. To output results continuously and not wait until the data streams end, blocking operators must be transformed into *non-blocking* operators. In fact, queries expressible by non-blocking operators are monotonic queries.

Definition 2.9 (*Monotonic query*) A continuous operator or continuous query Q is monotonic if $Q(\tau) \subseteq Q(\tau')$ for all $\tau \leq \tau'$.

Simple selection over a single stream is an example of monotonic query—at any point in time τ' when a new tuple arrives, it either satisfies selection predicate or it does not, and all the previously returned results (tuples) remain in $Q(\tau')$ [19]. Both standard and spatial aggregate operators always return a stream of length one—they are non-monotonic, and thus blocking.

Definition 2.10 (*Non-blocking query*) A non-blocking continuous operator or continuous query $Q_{\mathcal{NB}}$ is one that produces results (all the tuples of output) before it has detected the end of the input [31].

The problem of transforming the blocking queries into their non-blocking counterpart has long been recognized by data stream researchers. A dominant technique that overcomes this problem is to restrict the operator range to a finite *window* over input streams. Windows were introduced into standard SQL as part of SQL:1999 OLAP functions. In SQL:1999, window is a user-specified selection of rows within a query that determines the set of rows with respect to the current row under examination [33]. The motivation for having the window concept in DSMS is quite different. Window limits and focuses the scope of an operator or a query to a manageable portion of the data stream. A window is *stream-to-relation* operator [6]—it specifies a snapshot of a finite portion of a stream at any time point as a temporary *relation*. In other words, window transforms blocking operators and queries to compute in non-blocking manner. At the same time, the most recent data is emphasized, which is more relevant than the older data in the majority of data stream applications. There are several window types, though two basic types are being extensively used in conventional DSMS: logical, *time-based* windows and physical, *tuple-based* windows. By default, a time-based is refreshed at every time tick and tuple-based window is refreshed when a new tuple arrives. The tuples enter and expire from the window in a *first-in-first-expire* pattern: whenever a tuple becomes old enough, it is expired (i.e., deleted) from memory leaving its space to a more recent tuple. As a result, traditional window queries can support only (recent) historical queries, making them not suitable for spatio-temporal queries concerned with the current state of data rather than the recent history [37]. However, these two windows type are not useful in answering an interesting and important class of queries over spatio-temporal data stream, and the *predicate-based* window has been proposed [17], in which an arbitrary logical predicate specifies the window content.

2.2.1 Time-Based Window

The time-based window $\mathscr{W}_\omega^{time}$ is defined in terms of the window size ω represented as a time interval $\mathscr{T}_\omega$. Formally, it takes a stream $\mathscr{S}$ and the time interval $\mathscr{T}_\omega$ as parameters and returns finite, bounded stream, i.e. a temporary finite relation[5]:

$$\mathscr{W}_\omega^{time} : \mathscr{S} \times \mathscr{T}_\omega \to \mathscr{R}$$

The scope of time-based window $\mathscr{W}_\omega^{time}$ denotes the most recent time interval, i.e. it consists of the tuples whose timestamp is between the current time τ and $\tau - \omega$. Let τ_0 denotes the time instant that a continuous query specifying a sliding window has effectively started, and τ denotes time instant of the current time. The scope of time interval $\mathscr{T}_\omega$ may be formally specified as follows:

$$\mathscr{T}_\omega(\tau) = \begin{cases} [\tau_0, \tau] & \text{if } \tau_0 \leq \tau < \tau_0 + \omega \\ [\tau - \omega + 1, \tau] & \text{if } \tau \geq \tau_0 + \omega \end{cases}$$

The qualifying tuples are included in the window on the basis of their timestamps, by appending any newly tuples and discarding older ones. It is worth to note that time-based window is refreshed at every time instant, i.e., with constant refresh time granularity. We will use the following basic syntax to specify time-based window $\mathscr{W}_\omega^{time}$ on stream $\mathscr{S}$:

$\mathscr{S}$ [RANGE ω]

Example 2.1 Time-based window gpsStream [RANGE 20 seconds] defines a window over input stream gpsStream with the size of 20 s. At any time instant, the window output (relation) contains the bag of tuples from the previous 20 s.

There are two important subclasses of time-based window: *now* window $\mathscr{W}_{now}^{time}$ and *unbounded* window $\mathscr{W}_\infty^{time}$. Now window $\mathscr{W}_{now}^{time}$, defined by setting $\tau = \text{NOW}$ and $\omega = 1$, returns tuples of stream (relation) with timestamp equals to NOW. Unbounded window $\mathscr{W}_\infty^{time}$, defined by setting $\omega = \infty$ consists of tuples obtained from all tuples from stream $\mathscr{S}$ up to time instant τ. These two special windows are specified using the following syntax:

---

[5]The *instantaneous relation* in [6].

$\mathscr{S}$ [NOW]
$\mathscr{S}$ [RANGE UNBOUNDED]

Example 2.2 Suppose that Austrian rivers are stored in spatial database,[6] in `river` table created as:

```
CREATE TABLE river (
name :    VARCHAR(16),
geometry: GEOMETRY(POLYGON,3416)
)
```

Next query returns a position (type of `Point`) of moving objects that cross the Danube River at each time instant of the current time:

```
WITH danube AS (
     SELECT geometry FROM river WHERE name = 'Danube'
     )
SELECT position
FROM   movingObject [NOW]
WHERE  Crosses(movingObject.position, danube.geometry)
```

Example 2.3 Consider the following query:

```
WITH danube AS (
     SELECT geometry FROM river WHERE name = 'Danube'
     )
SELECT position
FROM   movingObject [RANGE UNBOUNDED]
WHERE  Crosses(movingObject.position, danube.geometry)
```

This query is monotonic, and produces relation that at time τ contains position of all moving objects that have crossed the Danube River up to τ.

When a stream is referenced in a query without a window specification, an unbounded window $\mathscr{W}_{\infty}^{time}$ is applied by default. Therefore, the next query is equivalent to the previous one:

```
WITH danube AS (
     SELECT geometry FROM river WHERE name = 'Danube'
     )
SELECT position
FROM   movingObject
WHERE  Crosses(movingObject.position, danube.geometry)
```

---

[6]We suppose that DSMS is coupled with spatial DBMS, but to preclude any potential transaction-processing issues that might occur concurrently with data stream processing, we will assume that the content of persistent relations involved in continuous query remain static [7].

Time-based window can optionally contain a *slide* parameter λ, indicating the granularity at which window slides, i.e. how frequently the window should be refreshed. Accordingly, we define the scope of time interval $\mathcal{T}_\omega$ as

$$
\mathcal{T}_\omega(\tau) = \begin{cases} [\tau_0, \tau] & \text{if } \tau_0 \leq \tau < \tau_0 + \omega \wedge (\tau - \tau_0) \, mod \, \lambda = 0 \\ [\tau - \omega + 1, \tau] & \text{if } \tau \geq \tau_0 + \omega \wedge (\tau - \tau_0) \, mod \, \lambda = 0 \\ \mathcal{T}_\omega(\tau - 1) & \text{if } (\tau - \tau_0) \, mod \, \lambda \neq 0, \end{cases}
$$

and use the following syntax construction for sliding time-based window $\mathcal{W}_{\omega,\lambda}^{time}$:

  $\mathcal{S}$ [RANGE ω SLIDE λ]

Example 2.4 The following query returns every minute the position of moving objects that have crossed the Danube River in the last 5 min:

```
WITH danube AS (
     SELECT geometry FROM river WHERE name = 'Danube'
     )
SELECT position
FROM   movingObject [RANGE '5 minutes' SLIDE '1 minute']
WHERE  Crosses(movingObject.position, danube.geometry)
```

2.2.2 Tuple-Based Window

The tuple-based window $\mathcal{W}^{tuple}$ defines its output stream over time by a window of the last N elements over its input stream. Formally, it takes a stream $\mathcal{S}$ and the natural number $N \in \mathbb{N}^*$ as parameters, and returns temporary finite relation:

  $\mathcal{W}_N^{tuple} : \mathcal{S} \times \mathbb{N}^* \rightarrow \mathcal{R}$

At any time instant τ, the output relation $\mathcal{R}$ consists of the N tuples of $\mathcal{S}$ with the largest timestamps $\leq\tau$ (or all tuples if the length of up to τ is $\leq$N). If more than one tuple has the same timestamp, we must choose one tuple in a non-deterministic way to ensure N tuples are returned. For this reason, tuple-based windows may be non-deterministic, and therefore may not be appropriate for streams in which timestamps are not unique.

We will use the following basic syntax to specify tuple-based window $\mathcal{W}^{tuple}$ on stream $\mathcal{S}$:

  $\mathcal{S}$ [ROWS N]

The special case of $N = \infty$ is specified by

  $\mathcal{S}$ [ROWS UNBOUNDED]

and is equivalent to time-based window

  $\mathcal{S}$ [RANGE UNBOUNDED]

Example 2.5 Tuple-based window gpsStream [ROWS 1] denotes the "latest" tuple in our gpsStream, which is very simple compared to reality. In reality, we will have a number of moving objects with the same timestamp, and the result of gpsStream [ROWS 1] will be ambiguous. As a result, the usability of tuple-based window in spatio-temporal applications is very limited.

2.2.3 Predicate-Based Window

An important issue in data stream query languages is the frequency by which the answer gets refreshed as well as the conditions that trigger the refresh. Coarser periodic refresh requirements are typically expressed as windows, but users in spatio-temporal applications may not be interested only in refreshing the query answer (i.e. window) in response to every tuple arrival. Consequently, a data stream query language should allow a user to express more general refresh conditions based on an arbitrary conditions: temporal, spatial, event, etc. For example, consider the following query:

Q^π: *Continuously, report the position of moving objects that cross the Danube River*.

At any time point τ, the *window-of-interest* for query Q^π includes only the moving objects that qualify predicate *"cross the Danube River"*. If a moving object O reports a position that crosses (i.e. is within) the Danube River, then it should be in Q^π's window. Whenever O reports position that disqualifies the predicate *"cross the Danube River"*, O expires from Q^π's window. It's important to note that objects enter and expire from Q^π's window in an *out-of-order* pattern. An object is expires (and hence is deleted) from Q^π's window only when the object reports another position that disqualifies the window predicate.

The semantics of time-based window query model reads as follows:

Q^{time}: *Continuously, report the position of moving objects that cross the Danube River in the last Ω time units*.

Where Ω is the size of time-based window. The query Q^{time} is semantically different from query Q^π: the window-of-interest in Q^π includes objects *"crossing the Danube River"* while the window-of-interest in Q^{time} includes objects that *"have crossed the Danube Rriver in the last Ω time units"*.

Definition 2.11 (*Predicate-based window query*) A predicate-based window query Q^π is defined over data stream $\mathcal{S}$ and window predicate Π over the tuples in $\mathcal{S}$. At any point in time τ the answer to Q^π equals the answer to snap-shot query Q_τ^σ, where Q_τ^σ is issued at time τ and the inputs to Q_τ^σ are the tuples in stream $\mathcal{S}$ that qualify the predicate Π at time τ.

We will use the following basic syntax to specify predicate-based window $\mathcal{W}^\pi$ on stream $\mathcal{S}$:

$\mathcal{S}\,[\Pi]$

where Π is the predicate that qualifies and disqualifies tuples into (and out of) the window.

Time-based and tuple-based window queries fail to answer some of predicate-based window queries. Let Q_∞^{time} denote the query in Example 2.3, which is *de facto* implementation of query Q^{time} with $\Omega = \infty$. The main difference between predicate-based window query Q^π

```
WITH danube AS (
    SELECT geometry FROM river WHERE name = 'Danube'
    )
SELECT position
FROM   movingObject AS mo
       [Crosses(mo.position, danube.geometry)]
```

and the query Q_∞^{time} in Example 2.3 is that a disqualified tuple in the predicate-based window may result in a negative tuple[7] as an output while a disqualified tuple in the WHERE clause predicate of Q_∞^{time} does not result in any output tuples. When a tuple t qualifies the WHERE predicate of Q_∞^{time} and is reported in answer, t will remain in the Q_∞^{time} query answer for ever. On the contrary, in the predicate-based window query model, when a tuple t qualifies the predicate Π and is reported in the Q^π answer, later, t may be deleted from the query answer if t receives an update so that t does not qualify the window predicate any more [17].

Similarly, the *now* window is semantically different from the predicate-based window. Thus semantics for the query with *now* window in Example 2.2 read as follows:

Q^{now}: *Report the positions of moving objects that cross the Danube River NOW.*

At any time, the answer of continuously running query Q^{now} will include only position of moving objects that cross the Danube River at time instant $\tau = NOW$. On the other hand, at any time instant τ the predicate-based window query Q^π may include positions of moving objects that have crossed the Danube River before τ.

2.3   OCEANUS—A Prototype of Spatio-Temporal DSMS

Most of the relevant research in the area of data streams has been done within several projects, each producing a prototype DSMS. Within the STREAM project [5, 6], streams are transformed into relations using window operators and then queried using standard relational operators. Results can then be transformed back into streams. The Aurora [2, 10] supports continuous queries (real-time processing), views, and ad hoc queries all using substantially the same mechanisms. The Aurora has been superseded by Borealis [1], a distributed multi-processor version of Aurora. The Stream Mill system [8, 30] developed at UCLA with the emphasis on data mining streaming data, enables the user to define custom aggregates which are then used to process streaming data. The Telegraph project at UC Berkley [11] explores adaptive

---

[7]In the pipelined query execution model with the negative tuples approach, a negative tuple is interpreted as a deletion of a previously produced positive tuple [6, 18].

dataflow architecture which enables it to make scheduling decisions for each tuple. The TelegraphCQ DSMS is based on the Telegraph framework and implemented as an extension of the PostgreSQL DBMS, with very limited spatial support. Work described in [26] represents a unification of two different SQL extensions for data streams and their associated semantics. A time-based execution provides a way to model simultaneity, while a tuple-based execution provides a way to react to primitive events as soon as they are seen by the system. The result is a new model that gives the user control over the granularity at which one can express simultaneity.

Research works and corresponding DSMS prototypes mentioned above have one thing in common: spatial and spatio-temporal data types and operations have been completely neglected.

The PLACE server [38] supports continuous query processing of spatio-temporal streams. It views spatial data streams as automatically changing relations incrementally calculating results and producing positive and negative updates of the result. It allows the user to construct complex queries out of simpler operators such as *inside* or *kNN* (k-nearest-neighbor). The scalable on-line execution (SOLE) algorithm [37] for continuous and on-line evaluation of concurrent continuous spatio-temporal queries over data streams, is implemented in the PLACE. This is one of the first attempts to furnish query processors in data stream management systems with the required operators and algorithms to support a scalable execution of concurrent continuous spatio-temporal queries over spatio-temporal data streams. The SOLE is a unified framework that deals with range queries as well as kNN queries. Furthermore, the SOLE implements a *predicate-based* window [17] especially suitable for spatio-temporal queries, because most of them are concerned with the current state of data rather than the recent history. Although the PLACE server, supported by the SOLE, allows the user to construct complex queries by encapsulating the spatio-temporal query algorithms into a very limited set of primitive spatio-temporal pipelined operators (e.g., *inside* and *kNN*), the native support for spatio-temporal data types and operations on them is missing. Opposed to that, our formally grounded framework natively supports a rich set of spatio-temporal data types and operations and is a significant step towards a full-fledged geospatial DSMS.

Research works presented in [40, 41] try to merge the moving objects and the data streams fields of research. TelegraphCQ is used to manage moving objects trajectories. The emphasis is placed on design and implementation of a powerful map application for managing, querying and visualizing the evolving locations of moving objects.

The GeoInsight framework [28] extends a commercial DSMS processing engine [35] in two directions. It integrates a spatial library to support the online processing of geospatial streaming data and implements an online analytical refinement and prediction (OARP) layer that enables querying of historical streaming data. These extensions allow users to issue various spatio-temporal queries about continuous events and enable the analysis of historical data, together with geospatial streaming data, to refine the answer of real-time queries and predict the answer in the near future. The extensibility infrastructure of the stream processing engine [35] provides user-defined aggregate (UDA), user-defined operator (UDO) and user-defined function

(UDF) facilities, which are used to seamlessly integrate spatial library and OARP modules into the query pipeline of the DSMS processing engine [3].

The work presented in [23] defines a geo-stream as a data stream carrying information about geometry or geometries changing over time by adding new data types called *stream* and *window*. It also defines their abstract semantics and operations on them. This work resolves two important issues: definition of windows semantics through a data type based approach and design of streaming data types, operations and predicates. It presents a novel approach in handling streaming data by encapsulating it inside a single attribute in a relation. This is one of the first works that formally and successfully merges geo-streams and moving objects.

OCEANUS formal framework [15, 16] may be most similar to that of Patroumpas and Sellis [41], Huang and Zhang [23], and Ali et al. [4, 28, 36] However, it overcomes limitations of [41], and is complementary to [4, 23, 28, 36] in the following way:

- Spatial data types in [41] are restricted to point objects only and do not include spatio-temporal operations. The scope of the model is limited to the movement of spatial point entities (i.e. spatial entities are considered to have no extent). The model also includes a restricted set of spatial predicates and operators proposed in an ad-hoc fashion, thus lacking in generality. The TelegraphCQ stream engine is used as a basis for the system, but with PostgreSQL built-in spatial operators and, therefore, with a rather limited and non-standard spatial support.
- OCEANUS is formally grounded in the framework of abstract spatio-temporal data types, it relies on relevant GIS standards [25] and full standardized spatial support for data types and operations, including a full set of spatio-temporal operations.
- The formal methodology taken in [23] is also based on many-sorted algebra, but considers *streams* and *windows* as data types. It includes a proof of the concept [46] as an extension of a moving objects database.
- OCEANUS strategy fully incorporates the data stream paradigm, viewing data streams as unbounded, explicitly timestamped and time-ordered sequences of tuples. It includes a prototype system implemented as a spatio-temporal extension of an object-relational DSMS.
- The approach taken in [4, 28, 36] and OCEANUS share a common feature: both rely on (albeit different) standard spatial library and support native approach to deal with spatial attributes as first-class citizens, to reason about the spatial properties of incoming events and to provide consistency guarantees over space and time. The work presented in [4, 28, 36] utilizes a generic approach to extend a streaming system for a particular application domain. This approach is focused on integrating user-defined modules (UDM) within the query pipeline of a DSMS [3]. Opposed to that, OCEANUS formal framework primarily focuses on formally extending and enriching a DSMS type system to support not only spatial, but also spatio-temporal attributes as first-class citizens. Additionally, it also includes a rich set of spatial, temporal and spatio-temporal operators.

2.3.1 The Type System

There are three kinds of different temporal predicates in spatio-temporal queries. Accordingly, the queries can be classified into three classes: historical, current and future queries [34]. In this section we define a system of data types on the abstract level supporting historical and current queries, using many-sorted algebra and second-order signature, building upon work in [12, 21, 23].

Modeling on the abstract level allows us to make definitions in terms of infinite sets, without fixing any finite representations of these sets. In the spatio-temporal context, a moving point is a continuous curve in 3D (2D + time) or 4D (3D + time) space, i.e. mapping from an infinite time domain into an infinite 2D (or 3D) space domain. The abstract level is the conceptual model that we are interested in, but we should keep in mind that this level of abstraction is not directly implementable and an additional step of choosing a concrete, finite representation is needed.

A many-sorted algebra [32] consists of sets and functions. A *signature* is a pair of sets (S, Ω), the elements of which are called *sorts* and *operations* respectively. Each operation consists of a (k+2)-tuple

$n : s_1 \times \cdots \times s_k \rightarrow s$, with $s_1, \ldots s_k, s \in S$ and $k \geq 0$.

Operation names (*operators*) are denoted by n and $s_1, \ldots, s_k, s \in S$ are sorts. In the case $k = 0$, the operation is called a *constant of sort s*. Informally, a sort denotes the name of a type, and an operation denotes a function [32]. A second-order signature [20] consists of two coupled many-sorted signatures:

1. The first signature defines a type system. In this context the sorts are called *kinds* and denote collections of types, and the *type constructors* have the role of operators. This signature generates a set of *terms*, which are exactly the *types* (Table 2.1).
2. The second signature defines a collection of polymorphic operations over the types of the type system.

The definitions of the *tuple* and *srelation* type constructors already use some extensions to the basic concepts of many-sorted algebra. The first extension is that a type constructor may use as sorts not only kinds (type collections), but also individual types. The second extension makes it possible to define operators taking a variable number of operands. The notation s^+ denotes a list of one or more operands of sort s. The third extension is that if $s_1, \ldots, s_n$ are sorts, then $(s_1 \times \cdots \times s_n)$ is also a sort.

The concepts of temporal types require additional consideration and will be described briefly. To model values of a spatial type σ that change over time, we introduce the notion of a temporal function which is an element of type

$\theta(\sigma) = \mathbb{T} \twoheadrightarrow \sigma$

The temporal functions are the basis of an algebraic model for the spatio-temporal data types, where σ is assigned one of the spatial data types *point, multipoint, linestring* or *polygon*, resulting in *tpoint* as values of type $\theta(point)$, *tmultipoint* as values of type $\theta(multipoint)$, *tlinestring* as values of type $\theta(linestring)$, *tmultilinestring*

Table 2.1 Type system—abstract model

Type constructor	Signature	Target sort
integer, real *string, boolean* *point, multipoint,* *linestring, multilinestring,*		$\rightarrow$ BASE
polygon		$\rightarrow$ SPATIAL
instant		$\rightarrow$ TIME
range	BASE $\cup$ TIME	$\rightarrow$ RANGE
temporal, intime	(BASE $\cup$ SPATIAL) $\times$ WINDOW	$\rightarrow$ TEMPORAL
tuple	((BASE $\cup$ SPATIAL $\cup$ TIME)$^+$ $\times$ TIME)	$\rightarrow$ TUPLE
stream	TUPLE	$\rightarrow$ STREAM
now, unbounded,		
range		$\rightarrow$ WINDOW
srelation	STREAM $\times$ WINDOW	$\rightarrow$ IRELATION

as values of type $\theta(multilinestring)$, and *tpolygon* as values of type $\theta(polygon)$. Consequently, we define base temporal types: *tinteger, tfloat, tboolean*, and *tinstant* which are relevant in the spatio-temporal context.

If we consider a situation where cars equipped with GNSS sensors are moving in a city, each towards a certain goal, *tinteger* could describe the number of cars within a certain area, *tboolean* could describe whether a certain area contains any cars at all, *tfloat* could describe the distance between a car and its goal, and *tinstant* could describe the estimated time of arrival of a car at its goal (which could change in time as traffic conditions vary).

Abstract semantics of BASE, SPATIAL, TIME, RANGE and TEMPORAL sorts have been precisely defined in [12, 13, 20–22]. BASE contains the atomic data types *int, real, bool* and *string*, while SPATIAL contains spatial data types *point, multipoint, linestring* and *polygon*. When applied to a compatible simple data type, temporal constructors produce a new data type e.g. *range(int), temporal(linestring)*, etc.

Abstract semantics of data type α are defined by its carrier set denoted by $\mathscr{C}_\alpha$. Each carrier set must contain an undefined value denoted by $\perp$. Type *instant* represents a point in time and is isomorphic to real numbers. The *range* constructor can be applied to BASE and TIME data types, has a starting and ending value and contains all values in between. The *temporal* constructors can be applied to BASE and SPATIAL data types (Table 2.1).

Table 2.2 Projection operations to domain and range

Operation	Signature	
deftime	$temporal(\alpha)$	$\rightarrow periods$
rangevalues	$temporal(\alpha)$	$\rightarrow range(\alpha)$
locations	$temporal(point)$	$\rightarrow multipoint$
	$temporal(multipoint)$	$\rightarrow multipoint$
trajectory	$temporal(point)$	$\rightarrow linestring$
	$temporal(multipoint)$	$\rightarrow multilinestring$
traversed	$temporal(line)$	$\rightarrow polygon$
	$temporal(polygon)$	$\rightarrow polygon$
instant	$intime(\alpha)$	$\rightarrow instant$
value	$temporal(\alpha)$	$\rightarrow \alpha$

2.4 Operators

As already stated in the previous section, the second signature defines a collection
of polymorphic operations over the types of the type system. The complete list of
polymorphic operations is rather extensive; a complete list of operators on the static
data types (*BASE, SPATIAL, TIME* and *RANGE*) is defined in [21, 22].[8]

All these static data types are made uniformly time dependent by introducing a
type constructor **temporal**. For a given static data type α, *temporal* returns the type
whose values are partial functions from the time domain $\mathbb{T}$ into α.

Let $\mathscr{D}_\alpha$ denote the domain of type α. The domain for type *temporal*(α) is

$$\mathscr{D}_{temporal(\alpha)} := \{f \mid f : \mathscr{D}_{instant} \nrightarrow \mathscr{D}_\alpha\}$$

The temporal types realized through the *temporal* type constructor are infinite
sets of pairs (instant, value), whereas the *intime* type constructor returns types representing single elements of such sets.

There are two classes of operation on temporal types. Projection class of operations
(Table 2.2) deals with projections of temporal values into domain and range, whereas
interaction class of operations (Table 2.3) relates the functional values of with values
in either their time or their range [22].

2.4.1 Lifting Operations to Spatio-Temporal Streaming Data Types

Operations on non-temporal types are then applied to *TEMPORAL* data types using
temporal. Temporal lifting is the key concept for achieving consistency of operations

---

[8]However, OCEANUS prototype, including operations on non-temporal types relies on the existing
operations and functions of TelegraphCQ [14] and PostGIS [43].

Table 2.3 Interaction operations with domain and range

Operation	Signature		Space dimension
atinstant	$temporal(\alpha) \times instant$	$\rightarrow intime(\alpha)$	
atperiods	$temporal(\alpha) \times periods$	$\rightarrow temporal(\alpha)$	
initial, final	$temporal(\alpha)$	$\rightarrow intime(\alpha)$	
present	$temporal(\alpha) \times instant$	$\rightarrow boolean$	
	$temporal(\alpha) \times periods$	$\rightarrow boolean$	
at	$temporal(\alpha) \times \beta$	$\rightarrow temporal(\alpha)$	1D
	$temporal(\alpha) \times \beta$	$\rightarrow temporal(min(\alpha, \beta))$	2D
atmin, atmax	$temporal(\alpha)$	$\rightarrow temporal(\alpha)$	1D
passes	$temporal(\alpha) \times \beta$	$\rightarrow boolean$	

on non-temporal and temporal types. All operations defined on static, non-temporal data types are systematically extended to corresponding temporal data types. The idea is to allow argument sorts and the target sort of a signature (the arguments and the result of the operation) to be of the temporal type. Consider the binary topological predicate ***inside*** for points and polygons that determines whether a point lies within a polygon:

$inside : point \times polygon \rightarrow boolean$

It is reasonable to expect that a similar operation can be calculated for a temporal point and a temporal polygon. However, since a temporal point can move in and out of a temporal polygon with time, the result of this lifted operation ***inside*** will also change with time. Thus, a lifted operation ***inside*** will have the following signatures:

$inside : temporal(point) \times polygon \qquad\qquad \rightarrow temporal(boolean)$

$inside : point \times temporal(polygon) \qquad\qquad \rightarrow temporal(boolean)$

$inside : temporal(point) \times temporal(polygon) \quad \rightarrow temporal(boolean)$

If we abbreviate the formally defined notation $temporal(\alpha)$ with $t\alpha$, a lifted operation ***inside*** has the following signatures:

$inside : tpoint \times polygon \quad \rightarrow tboolean$

$inside : point \times tpolygon \quad \rightarrow tboolean$

$inside : tpoint \times tpolygon \rightarrow tboolean$

The predicate yields true for each time instant τ at which the temporal point is inside the polygon, undefined whenever the point or the polygon is undefined, and false otherwise.

Formally, for each operation with a signature

$o : s_1 \times \cdots \times s_n \rightarrow s$

its corresponding temporally lifted version is defined by:

$\Uparrow o : \theta(s_1) \times \cdots \times \theta(s_n) \rightarrow \theta(s)$

More specifically, OCEANUS collection of spatio-temporal polymorphic operations consists of temporally lifted versions of spatial operations defined in [25, 43].

2.5 Implementation

A discrete model, as a finite instantiation of an abstract model, is needed for implementation. The discrete model uses the so-called sliced representation: temporal changes of a value along the time dimension are decomposed into fragment intervals called slices. The temporal changes for temporal points and temporal polygons can be represented within each slice by a simple linear function. Figure 2.3 shows a single slice of a polygon changing in time (*tpolygon*) using linear interpolation between subsequent timestamps of two data stream tuples. The design of efficient algorithms for the operations of the abstract model is based on the premise that data structures for different data types are intended to be used within a DSMS, which implies that the data will be placed into the main memory under the control of the DSMS.

In the discrete model, the constructor *temporal* is replaced by type constructors that enable the so-called sliced representation of temporal types (Table 2.4).

Definition 2.12 (*Generic Unit Type*) Let S be a set.
$Unit(S) = Interval(Instant) \times S$
A pair (i, v) from $Unit(S)$ is called a *unit*, i is its *unit interval*, and v is called its *unit function*.

The semantics of a unit type are a function of time, defined during the unit interval:
$\iota : S_\alpha \times Instant \rightarrow \mathscr{D}_\alpha$
where α is the corresponding non-temporal type.

Definition 2.13 (*Mapping Type Constructor*) Let S be a set and $Unit(S)$ a unit type.

$$Mapping(S) = \{U \subset Unit(S) \mid \forall (i_1, v_1) \in U, (i_2, v_2) \in U :$$
$$i_1 = i_2 \Rightarrow v_1 = v_2,$$
$$i_1 \neq i_2 \Rightarrow (disjoint(i_1, i_2) \wedge adjacent(i_1, i_2) \Rightarrow v_1 \neq v_2)\}$$

The type constructor *mapping* represents a set of unit types whose values are pairs consisting of a time interval and a simple function, describing temporal development

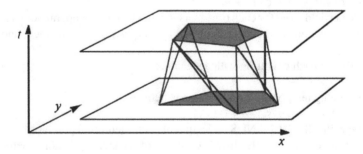

Fig. 2.3 A temporal polygon unit (*source* [22])

Table 2.4 Type system—discrete model

Type constructor	Signature	Target sort
…	…	…
const	$(\text{BASE} \cup \text{SPATIAL}) \times \text{WINDOW}$	$\rightarrow$ UNIT
upoint, umultipoint,		
ulinestring, umultilinestring,		
upolygon		$\rightarrow$ UNIT
mapping	UNIT	$\rightarrow$ TEMPORAL
…	…	…

of a value during that time interval. A mapping is a set of units whose time intervals are pairwise disjoint.

The spatial attribute $\mathscr{A}_\sigma$ and temporal attribute $\mathscr{A}_\theta$ represent two related but separate attribute values. Together they represent a unique *spatio-temporal reference* of a data stream element. They are implemented with composite temporal data type, whose values are pairs of time instant τ (the value of $\mathscr{A}_\theta$) and a shape/location (the value of $\mathscr{A}_\sigma$). This data type is derived using the type constructor *intime* in a data acquisition process that manages connections with external data sources and the functions which read data from these data sources. The time instant[9] component implicitly used in the type constructor *intime* represents valid time.

Definition 2.14 (*Spatio-Temporal Reference*) A spatio-temporal reference of a data stream element is obtained by applying the type constructor *intime* to an appropriate spatial data type, i.e. *intime* (σ).

The carrier set for *intime*(σ) at the discrete level is

$$\mathscr{D}_{intime(\sigma)} := \mathscr{D}_{instant} \times \mathscr{D}_\sigma$$

2.5.1 User-Defined Aggregate Functions

The set of aggregate functions available to users of a data stream system is usually limited to five standard functions: *min, max, count, sum, and avg*. Over time it became apparent that users often want to aggregate data in additional ways. Therefore, many data stream systems allow the user to extend the set of available aggregate functions by defining user-defined aggregate functions (UDAF).

---

[9]In OCEANUS prototype, the *instant* is implemented as PostgreSQL timestamp data type.

To create a UDAF, the user must implement at least three functions[10]:

Init   This function is used to initialize any variables needed for the com-
              putation later on. Intuitively, it is similar to constructor.
Accumulate   This function is called once for each value aggregated. Generally, this
              function "adds" the value to the running total computed so far.
Terminate   This function is used to end the calculation and return the final value
              of the aggregate function. It may involve some calculations on the
              variables which were defined for use with the aggregate function.

These required functions can be implemented in a external procedural language, but they can also be specified natively using a query language supported by the DSMS. The second approach is less expressive, but enables these three computations to be coded in one single procedure written in SQL, rather than in three separate procedures using an external procedural language. This approach is used in the Expressive Stream Language (ESL), an application language of the Stream Mill system [45].

Similar to continuous queries, a UDAF is considered as *blocking* [19] if it requires its entire input before it can return a result (i.e. it must have a terminate operation), or if, for computing the aggregate result, the data first has to be sorted according to GROUP BY attributes. Such blocking aggregates can only be applied over data streams using a window operator and are termed *windowed aggregates*, while UDAFs invoked without a window operator are termed *base aggregates*. UDAFs in which terminate operation is not defined are called non-blocking and can be applied to data streams freely.

The proposed temporal data type for presenting a spatio-temporal reference of a data stream element enables the implementation of a sliced representation of spatio-temporal objects. A spatio-temporal object is formed of unit types, each consisting of a time interval and a value describing a simple function on the unit time interval. If we consider, for example, two successive spatio-temporal data stream elements with spatio-temporal references consisting of successive point locations and time instants, it is possible to define the unit type that describes the movement of the corresponding point object as a linear function of time.

The complete sliced representation presents a temporal object as a set of temporal units whose time intervals are disjoint, and if adjacent, their values are different. The assembling of the unit types into a temporal object based on the spatio-temporal references of data stream elements can be achieved using UDAF. We have already seen that aggregation can be defined as a state that is modified with every input data. In terms of spatio-temporal data streams, such a state represents a temporal object in a form of described sliced representation (*mapping(upoint)* data type), while a spatio-temporal reference of a data stream element represents new input data (*intime(point)*

---

[10]The exact names of the functions (and the class to which they must belong) differs depending on the concrete DSMS.

data type). Function `accumulate` that is used in UDAFs, performs the creation of a unit type and adds it to the temporal object. Since this function computes a new state (temporal object) from the old one and the value of the spatio-temporal reference of a new data stream element, the UDAF does not need a terminate function and is thus considered non-blocking. This concept can be implemented using the following PostgreSQL-like syntax:

```
CREATE AGGREGATE tpoint(intime(point))
       SFUNC = accumulate,
       STYPE = mapping(upoint),
       INITCOND = NULL
);
```

An aggregate function is made from one or two ordinary functions: a state transition (*accumulate*) function `SFUNC`, and an optional *terminate* function `FINALFUNC`. Since UDAF for extracting a moving object from a data stream is non-blocking, the terminate function `FINALFUNC` is omitted. A temporary variable of data type `STYPE` holds the current internal state of the aggregate. At each input row, the aggregate argument value(s) are calculated and the state transition function is invoked with the current state value and the new argument value(s) in order to calculate a new internal state value. In the UDAF definition given above, the name of the function is `tpoint`, the data type for storing internal state and the result of the function is `mapping(upoint)` (discrete data type for temporal points) and the data type for the new input data is `intime(point)` (given as an argument of the defined UDAF). An aggregate function provides an initial condition (`INITCOND`), that is, an initial value for the internal state. A spatio-temporal object is not defined before the first input stream tuple has arrived, and the initial value for the internal state is NULL. After all the stream tuples rows have been processed, the final function is invoked once to calculate the aggregates return value. If there is no final function, then the ending state value is returned unchanged [39].

Algorithms 1 and 2 describe an abstract implementation of UDAF `tpoint` and `accumulate` functions respectively.

Algorithm 1 *tpoint(IP)*

---

input: set *IP* of spatio-temporal references of type *ipoint*
output: temporal point object *tp* of STYPE *tpoint*
summary: A set of spatio-temporal references is aggregated into a temporal point object.
1: INITCOND ← NULL
2: **for all** *ip* ∈ *IP* **do**
3:     *tp* ← *accumulate*(*tp*, *ip*)
4: **end for**
5: **return** *tp*

Algorithm 2 *accumulate(tp,ip)*

---

input: temporal point object *tp* of type *tpoint*, ST reference *ip* of type *ipoint*
output: temporal point object *tp* of type *tpoint*
summary: Temporal point object *tp* consists of units. Each unit consists of a time interval and a linear function valid during that time interval. The algorithm takes a new ST reference *ip* and tries to add it to the last unit in the moving object *mo*. If that is not possible, it creates a new unit using the new ST reference *ip* and the ending ST reference of the last unit object *ip_l*. It then adds the new unit to the moving object.

1: **if** $tp = \emptyset$ **then**
2:     $fun.x_0 \leftarrow ip.\sigma.x, fun.x_1 \leftarrow 0$ ▷ *fun* - linear function valid during time interval
3:     $fun.y_0 \leftarrow ip.\sigma.y, fun.y_1 \leftarrow 0$
4:     $ti.start \leftarrow ti.end \leftarrow ip.\theta$ ▷ *tp* - time interval
5: **else**
6:     $up \leftarrow tp.units.last$
7:     $ip_l \leftarrow up.endpoint$
                        ▷ linear movement function $= f((x_0, x_1, y_0, y_1), t) = (x_0 + x_1 t, y_0 + y_1 t)$
8:     $fun.x_0 \leftarrow (-ip.\theta) * (ip_l.\sigma.x - ip.\sigma.x)/$
    $(ip_l.\theta - ip.\theta) + ip_l.\sigma.x$
9:     $fun.x_1 \leftarrow (ip_l.\sigma.x - ip.\sigma.x)/(ip_l.\theta - ip.\theta)$
10:     $fun.y_0 \leftarrow (-ip.\theta) * (ip_l.\sigma.y - ip.\sigma.y)/$
    $(ip_l.\theta - ip.\theta) + ip_l.\sigma.y$
11:     $fun.y_1 \leftarrow (ip_l.\sigma.y - ip.\sigma.y)/(ip_l.\theta - ip.\theta)$
12:     **if** $fun = up.function$ **then**
13:         $up.timeinterval.end \leftarrow ip.\theta$
14:         **return** tp
15:     **else**
16:         $ti.start \leftarrow ip_l.\theta$
17:         $ti.end \leftarrow ip.\theta$
18:     **end if**
19: **end if**
20: create new unit object up_{new}
21: $up_{new}.timeinterval \leftarrow ti$
22: $up_{new}.function \leftarrow fun$
23: $tp.add(up_{new})$
24: **return** tp

In combination with different window operators, i.e. stream-to-relation operators [6] on spatio-temporal data streams, the proposed UDAF constructs temporal point object in a form of a sliced representation as intermediate results of a continuous query.

2.5.2 SQL-Like Language Embedding: CSQL

The proposed concept is illustrated through CSQL, an SQL-like query language that supports continuous queries over spatio-temporal data streams. In the following examples we show how the query language is extended with spatio-temporal data types and a set of operations over those data types such as *passes, deftime, _value_, trajectory*, etc. Our examples are based on the following spatio-temporal data stream schema:

```
carStream (
    id:        INTEGER,
    carType:   VARCHAR(8),
    driver:    SMALLINT,
    geometry:  intime(point),
    speed:     REAL,
    tcqtime:   TIMESTAMP TIMESTAMPCOLUMN
)
```

Attribute tcqtime represents valid time instant of a stream tuple which obeys the timestampcolumn constraint, which is assumed to be monotonically increasing.

The following examples illustrate spatio-temporal continuous queries and their visualization (Fig. 2.4) enabled within OCEANUS prototype:

Fig. 2.4 OCEANUS interface—visualisation of *Q8*

Q1: *Find all cars within the area of interest (e.g. a rectangle).*

```
SELECT id
FROM    carStream [NOW]
GROUP   BY id
HAVING  passes(tpoint(geometry),
                GeometryFromText
                ('POLYGON(575000.0 322000.0, 557000.0 323000.0)',
                 4326)
                )
```

Q2: *For each car, find its minimal distance from the point of interest during the last half an hour.*

```
SELECT id, _value_(initial(atmin(distance(tpoint(geometry),
                SetSRID(POINT(557234, 322701), 4326)))))
FROM    carStream [RANGE 30 minutes]
GROUP BY id
```

Q3: *Find each time instant when car 1007 was heading north (let us say, within* 10° *of the exact north direction) during the last ten minutes.*

```
SELECT deftime(at(tdirection(tpoint(geometry)), range(80, 100)))
FROM    carStream [RANGE 10 minutes]
WHERE   car = 1007
```

Q4: *Find all cars/drivers that have travelled more than* 50 km *during the last hour.*

```
SELECT id, driver
FROM    carStream [RANGE 1 hour]
GROUP   BY id, driver
HAVING  length(trajectory(tpoint(geometry))) > 50000
```

Q5: *Report the trajectories of cars with id 1, 2, 3 and 4 in the last* 2 h.

```
SELECT id, trajectory(tpoint(geometry))
FROM    carStream [RANGE 2 hours]
WHERE   id IN (1, 2, 3, 4)
GROUP   BY id
```

Q6: *Find trajectories of the cars that were within* 100 m *from a point of interest within the last* 30 min.

```
SELECT id, trajectory(tpoint(geometry))
FROM    carStream [RANGE 30 minutes]
GROUP   BY id
HAVING  _value_(initial(atmin(distance(tpoint(geometry),
        SetSRID(POINT(557234, 322701), 4326)))))) <= 100
```

Q7: *Report the movement of all cars that have crossed the street number 10 in the last* 2 h.

```
SELECT id, trajectory(tpoint(carStream.geometry))
FROM   carStream [RANGE 2 hours], streetTable
WHERE  streetid = 10
GROUP  BY id
HAVING intersects(trajectory(tpoint(carStream.geometry)),
                  streetTable.geometry)
```

Q8: *Continuously report the location of all cars within a particular city district.*

```
WITH district AS (
     SELECT geometry FROM cityDistrict WHERE name = 'Crnomerec'
     )
SELECT id, speed, carStream.geometry AS location
FROM   carStream [NOW]
WHERE  Within(carStream.geometry, district.geometry)
```

References

1. Abadi, D.J., Ahmad, Y., Balazinska, M., Çetintemel, U., Cherniack, M., Hwang, J., Lindner, W., Maskey, A., Rasin, A., Ryvkina, E., Tatbul, N., Xing, Y., Zdonik, S.B.: The design of the Borealis stream processing engine. In: CIDR, pp. 277–289 (2005). http://www.cidrdb.org/cidr2005/papers/P23.pdf
2. Abadi, D.J., Carney, D., Çetintemel, U., Cherniack, M., Convey, C., Lee, S., Stonebraker, M., Tatbul, N., Zdonik, S.B.: Aurora: a new model and architecture for data stream management. Int. J. Very Large Databases **12**(2), 120–139 (2003)
3. Ali, M.H., Chandramouli, B., Goldstein, J., Schindlauer, R.: The extensibility framework in Microsoft StreamInsight. In: Abiteboul, S., Böhm, K., Koch, C., Tan, K. (eds.) Proceedings of the 27th International Conference on Data Engineering, ICDE 2011, pp. 1242–1253. IEEE Computer Society (2011)
4. Ali, M.H., Chandramouli, B., Raman, B.S., Katibah, E.: Spatio-temporal stream processing in Microsoft StreamInsight. IEEE Data Eng. Bull. **33**(2), 69–74 (2010)
5. Arasu, A., Babcock, B., Babu, S., Datar, M., Ito, K., Motwani, R., Nishizawa, I., Srivastava, U., Thomas, D., Varma, R., Widom, J.: STREAM: the stanford stream data manager. IEEE Data Eng. Bull. **26**(1), 19–26 (2003)
6. Arasu, A., Babu, S., Widom, J.: The CQL continuous query language: semantic foundations and query execution. Int. J. Very Large Databases **15**(2), 121–142 (2006)
7. Babcock, B., Babu, S., Datar, M., Motwani, R., Widom, J.: Models and issues in data stream systems. In: Popa, L., Abiteboul, S., Kolaitis, P.G. (eds.) PODS, pp. 1–16. ACM (2002)
8. Bai, Y., Thakkar, H., Wang, H., Luo, C., Zaniolo, C.: A data stream language and system designed for power and extensibility. In: Yu, P.S., Tsotras, V.J., Fox, E.A., Liu, B. (eds.) CIKM, pp. 337–346. ACM (2006)
9. Bettini, C., Dyreson, C.E., Evans, W.S., Snodgrass, R.T., Wang, X.S.: A glossary of time granularity concepts. In: Temporal Databases, Dagstuhl, pp. 406–413 (1997)
10. Carney, D., Çetintemel, U., Cherniack, M., Convey, C., Lee, S., Seidman, G., Stonebraker, M., Tatbul, N., Zdonik, S.B.: Monitoring streams - a new class of data management applications. In: VLDB, pp. 215–226. Morgan Kaufmann (2002)

11. Chandrasekaran, S., Cooper, O., Deshpande, A., Franklin, M.J., Hellerstein, J.M., Hong, W., Krishnamurthy, S., Madden, S., Reiss, F., Shah, M.A.: TelegraphCQ: continuous dataflow processing. In: Halevy, A.Y., Ives, Z.G., Doan, A. (eds.) SIGMOD Conference, p. 668. ACM (2003)
12. Erwig, M., Güting, R.H., Schneider, M., Vazirgiannis, M.: Abstract and discrete modeling of spatio-temporal data types. In: Laurini, R., Makki, K., Pissinou, N. (eds.) ACM-GIS '98, Proceedings of the 6th International Symposium on Advances in Geographic Information Systems, November 6–7, 1998, Washington, DC, USA, pp. 131–136. ACM (1998). http://doi.acm.org/10.1145/288692.288716
13. Forlizzi, L., Güting, R.H., Nardelli, E., Schneider, M.: A data model and data structures for moving objects databases. In: Chen, W., Naughton, J.F., Bernstein, P.A. (eds.) Proceedings of the 2000 ACM SIGMOD International Conference on Management of Data, May 16–18, 2000, Dallas, Texas, USA, pp. 319–330. ACM (2000). http://doi.acm.org/10.1145/342009.335426
14. Franklin, M.J., Krishnamurthy, S., Conway, N., Li, A., Russakovsky, A., Thombre, N.: Continuous analytics: Rethinking query processing in a network-effect world. In: CIDR (2009). www.cidrdb.org
15. Galić, Z., Baranović, M., Križanović, K., Mešković, E.: Geospatial data streams: Formal framework and implementation. Data Knowl. Eng. **91**, 1–16 (2014). http://dx.doi.org/10.1016/j.datak.2014.02.002
16. Galić, Z., Mešković, E., Križanović, K., Baranović, M.: Oceanus: a spatio-temporal data stream system prototype. In: Proceedings of the Third ACM SIGSPATIAL International Workshop on GeoStreaming, pp. 109–115. IWGS '12, ACM, New York, NY, USA (2012). http://doi.acm.org/10.1145/2442968.2442982
17. Ghanem, T.M., Aref, W.G., Elmagarmid, A.K.: Exploiting predicate-window semantics over data streams. SIGMOD Rec. **35**(1), 3–8 (2006)
18. Ghanem, T.M., Hammad, M.A., Mokbel, M.F., Aref, W.G., Elmagarmid, A.K.: Incremental evaluation of sliding-window queries over data streams. IEEE Trans. Knowl. Data Eng. **19**(1), 57–72 (2007)
19. Golab, L., Özsu, M.T.: Data Stream Management. Synthesis Lectures on Data Management. Morgan Claypool Publishers, San Rafael (2010)
20. Güting, R.H.: Second-order signature: a tool for specifying data models, query processing, and optimization. In: Buneman, P., Jajodia, S. (eds.) SIGMOD Conference, pp. 277–286. ACM Press (1993)
21. Güting, R.H., Böhlen, M.H., Erwig, M., Jensen, C.S., Lorentzos, N.A., Schneider, M., Vazirgiannis, M.: A foundation for representing and quering moving objects. ACM Trans. Database Syst. **25**(1), 1–42 (2000)
22. Güting, R.H., Schneider, M.: Moving Objects Databases. Morgan Kaufmann, Amsterdam (2005)
23. Huang, Y., Zhang, C.: New data types and operations to support geo-streams. In: Cova, T.J., Miller, H.J., Beard, K., Frank, A.U., Goodchild, M.F. (eds.) GIScience. Lecture Notes in Computer Science, vol. 5266, pp. 106–118. Springer (2008)
24. ISO 19107:2003: Geographic information – Spatial schema (2008)
25. ISO/IEC 13249-3:2011: Information technology – Database languages – SQL multimedia and application packages – Part 3: Spatial (2011)
26. Jain, N., Mishra, S., Srinivasan, A., Gehrke, J., Widom, J., Balakrishnan, H., Çetintemel, U., Cherniack, M., Tibbetts, R., Zdonik, S.B.: Towards a streaming SQL standard. Proc. VLDB Endow. **1**(2), 1379–1390 (2008)
27. Jensen, C.S., Dyreson, C.E., Böhlen, M.H., Clifford, J., Elmasri, R., Gadia, S.K., Grandi, F., Hayes, P.J., Jajodia, S., Käfer, W., Kline, N., Lorentzos, N.A., Mitsopoulos, Y.G., Montanari, A., Nonen, D.A., Peressi, E., Pernici, B., Roddick, J.F., Sarda, N.L., Scalas, M.R., Segev, A., Snodgrass, R.T., Soo, M.D., Tansel, A.U., Tiberio, P., Wiederhold, G.: The consensus glossary of temporal database concepts - february 1998 version. In: Temporal Databases, Dagstuhl, pp. 367–405 (1997)

28. Kazemitabar, S.J., Demiryurek, U., Ali, M.H., Akdogan, A., Shahabi, C.: Geospatial stream query processing using Microsoft SQL Server StreamInsight. PVLDB **3**(2), 1537–1540 (2010)
29. Krämer, J., Seeger, B.: Semantics and implementation of continuous sliding window queries over data streams. ACM Trans. Database Syst. 34(1) (2009)
30. Law, Y.N., Wang, H., Zaniolo, C.: Query languages and data models for database sequences and data streams. In: Nascimento, M.A., Özsu, M.T., Kossmann, D., Miller, R.J., Blakeley, J.A., Schiefer, K.B. (eds.) VLDB, pp. 492–503. Morgan Kaufmann (2004)
31. Law, Y.N., Wang, H., Zaniolo, C.: Relational languages and data models for continuous queries on sequences and data streams. ACM Trans. Database Syst. **36**(2), 8:1–8:32 (2011)
32. Loeckx, J., Ehrich, H.D., Wolf, M.: Specification of Abstract Data Types. Wiley and B. G Teubner (1996)
33. Melton, J.: Advanced SQL 1999: Understanding Object-Relational, and Other Advanced Features. Elsevier Science Inc., New York (2003)
34. Meng, X., Chen, J.: Moving Objects Management: Models, Techniques and Applications. Tsinghua University Press and Springer (2010)
35. Microsoft: Microsoft StreamInsight (2013). http://msdn.microsoft.com/en-us/library/hh750618(v=sql.10).aspx
36. Miller, J., Raymond, M., Archer, J., Adem, S., Hansel, L., Konda, S., Luti, M., Zhao, Y., Teredesai, A., Ali, M.H.: An extensibility approach for spatio-temporal stream processing using Microsoft StreamInsight. In: Pfoser, D., Tao, Y., Mouratidis, K., Nascimento, M.A., Mokbel, M.F., Shekhar, S., Huang, Y. (eds.) SSTD. Lecture Notes in Computer Science, vol. 6849, pp. 496–501. Springer (2011)
37. Mokbel, M.F., Aref, W.G.: SOLE: scalable on-line execution of continuous queries on spatio-temporal data streams. Int. J. Very Large Databases **17**(5), 971–995 (2008)
38. Mokbel, M.F., Xiong, X., Hammad, M.A., Aref, W.G.: Continuous query processing of spatio-temporal data streams in PLACE. GeoInformatica **9**(4), 343–365 (2005)
39. Obe, R., Hsu, L.: PostgreSQL - Up and Running: a Practical Guide to the Advanced Open Source Database. O'Reilly (2012)
40. Patroumpas, K., Kefallinou, E., Sellis, T.K.: Monitoring continuous queries over streaming locations. In: Aref, W.G., Mokbel, M.F., Schneider, M. (eds.) GIS, p. 81. ACM (2008)
41. Patroumpas, K., Sellis, T.K.: Managing trajectories of moving objects as data streams. In: Sander, J., Nascimento, M.A. (eds.) STDBM, pp. 41–48 (2004)
42. Patroumpas, K., Sellis, T.K.: Maintaining consistent results of continuous queries under diverse window specifications. Inf. Syst. **36**(1), 42–61 (2011)
43. Refractions Research Inc.: PostGIS Manual (2015)
44. Schneider, M.: Spatial Data Types for Database Systems, Finite Resolution Geometry for Geographic Information Systems, Lecture Notes in Computer Science, vol. 1288. Springer (1997)
45. Thakkar, H., Zaniolo, C.: Introducing Stream Mill: User-Guide to the Data Stream Management System, its Expressive Stream Language ESL, and the Data Stream Mining Workbench SMM. Computer Science Department, UCLA (2010)
46. Zhang, C., Huang, Y., Griffin, T.: Querying geospatial data streams in SECONDO. In: Agrawal, D., Aref, W.G., Lu, C., Mokbel, M.F., Scheuermann, P., Shahabi, C., Wolfson, O. (eds.) 17th ACM SIGSPATIAL International Symposium on Advances in Geographic Information Systems, ACM-GIS 2009, November 4–6, 2009, Seattle, Washington, USA, Proceedings, pp. 544–545. ACM (2009). http://doi.acm.org/10.1145/1653771.1653868

Chapter 3
Spatio-Temporal Data Streams and Big Data Paradigm

Abstract Recent rapid development of wireless communication, mobile computing, global navigational satellite systems (GNSS), and spatially enabled sensors is leading to an exponential growth of available spatio-temporal data produced continuously at hight speed. Spatio-temporal data streams, i.e. real-time, transient, time-varying sequences of spatiotemporal data items, demonstrates at least two Big Data core features: *volume* and *velocity*. To handle the volumes of data and computation they involve, these applications need to be distributed over clusters. However, despite substantial work on cluster programming models for batch computation, there are few similarly high-level tools for stream processing. Obviously, there is a clear need for highly scalable spatio-temporal stream computing framework that can operate at high data rates and process massive amounts of big spatio-temporal data streams. In this chapter we present our approach and framework for an integrated big spatio-temporal data stream processing. The key concept here is that streaming data and persistent data are not intrinsically different - the persistent spatio-temporal data is simply streaming data that has been entered into the persistent structures.

Keywords Big data · Data stream architectures · GeoStreaming · Mobility data · Parallel processing · Real-time distributed · Spatio-temporal data streams

3.1 Background

Big Data is a term used to identify the data sets that, due to their large size, we cannot manage without using new technologies and programming frameworks. It does not refer to any specific quantity, but rather to high-volume, high-velocity, high-variety and high-veracity data that demand cost-effective, innovative forms of data processing.

The distributed computing field has achieved success in scaling up big data processing on large numbers of unreliable, commodity machines. More recently, the Big Data paradigm has resulted in cluster computing frameworks for storage and large scale processing of huge data sets, the most prominent one being MapReduce [21] and Hadoop [58] with HiveQL [16] as a companion query language. Unfortu-

© The Author(s) 2016

Z. Galić, *Spatio-Temporal Data Streams*, SpringerBriefs
in Computer Science, DOI 10.1007/978-1-4939-6575-5_3

nately, they have not been truly exploited towards processing big spatio-temporal data streams.

A number of DBMS vendors also offers data stream processing engines, but only few of them support spatial data types and operations either by including geospatial toolkit [13], spatial cartridge [50], or extensibility framework to invoke methods from a spatial library [45], without distributed cluster computing capabilities.

Fluctuations in the data volume and the requirements for real-time responses to posed spatio-temporal queries, have been indicated as serious and perhaps even insurmountable obstacles both for the traditional spatio-temporal databases and cluster computing frameworks. The obstacles are primarily the consequence of *store-and-then-query* data processing paradigm, i.e., data are stored in the database/file system and *ad hoc* queries are answered in full, based on the current snapshot of data. The mismatch between high latency of secondary storage and low latency of main memory adds considerable delay in response time that is not acceptable to many monitoring applications. It is obvious that IFP applications do not readily fit neither DBMS nor MapReduce/Hadoop/Hive processing model due to their *outbound (pull-based)* processing paradigm: dynamic, transient *ad hoc* queries are typically specified, optimized and processed once over static, persistent data. Real-time processing of big spatio-temporal data streams has an essentially different set of requirements than batch processing, and requires complementary *inbound (push-based)* paradigm: the same, static, persistent queries are processed continuously over transient, dynamic, frequently changing streaming data.

Even though MapReduce framework, Hadoop, and related technologies have made it possible to store and process large-scale data, they are not real-time computing systems (RTCS). MapReduce OnLine [20]—the pipelined version of Hadoop, allows for near-real-time analysis of data streams, and thus allows the MapReduce programming model to be applied to IFP application domains.

Early streaming systems are focused on relational style operators as computations, whereas current systems support more general user-defined computations. Stream processing systems commonly define computation over streams as *work-flow*—a directed acyclic graph (DAG), in which nodes represent streaming operators in form of user-defined functions (UDFs), and edges represent an execution ordering.

Apache Storm [8] is a distributed real-time computation system for processing large volumes of high-velocity data. S4 (Simple Scalable Streaming System) [9, 48] is a general-purpose, distributed, scalable, fault-tolerant, pluggable platform that allows programmers to easily develop applications for processing continuous, unbounded streams of data. Storm natively supports primitive types, strings, and byte arrays as string tuple filled values, whereas data events in S4 could be any serializable Java objects. Both, Storm and S4, are based on a *record-at-a-time* processing model, where nodes receive each record, update internal state, and send out new records in response. This model raises several challenges in a large-scale cloud environment, including fault tolerance, limited scalability, consistency and unification with bah processing.

Apache Spark [5] is a cluster computing platform which improves over Hadoop MapReduce in several dimensions. By extending and generalizing MapReduce framework it goes far beyond batch applications and enable combining multiple types of computation (SQL queries, streaming, machine learning, etc.) that might previously have required different Big Data engines.

TimeStream [53] adopts the programming model of StreamInsight [3] for complex event processing (CEP), and extends it to large-scale distributed execution by providing automatic supports for parallel execution, fault tolerance, and dynamic reconfiguration. Similarly to DSMS, these cluster computing platforms do not possess spatio-temporal capabilities.

Naiad [47] is a distributed system for executing data parallel, cyclic dataflow programs. It offers the high throughput of batch processors, the low latency of stream processors, and the ability to perform iterative and incremental computations. High-level programming models can be built on Naiad's low-level primitives, enabling such diverse tasks as streaming data analysis, iterative machine learning, and interactive graph mining.

Trill [17] is a new query processor for analytics that fulfills a combination of three requirements for a query processor to serve the diverse big data analytics space: query model, fabric and language integration, and performance. It uses a streaming batched-columnar data representation with a new dynamic compilation-based system architecture that addresses all these requirements. Trill is based on the temporal logical data model, which enables the diverse spectrum of analytics: real-time, offline, temporal, relational, and progressive.

S-Store [44] addresses the state management shortcomings of many stream processing systems. In particular, it incorporates ACID transactions by building on H-Store, a main-memory OLTP DBMS.

Similarly to centralized DSMS, these distributed computing platforms do not possess spatio-temporal capabilities—a *conditio sine qua non* for big mobility data processing.

Spatio-temporal stream processing in general refers to processing of high volume spatio-temporal data streams with very low latency, i.e. in near real-time. However, all general-purpose, distributed stream processing engines [5–8, 12, 53] are missing support for real-time processing of large-scale spatio-temporal mobility data.

Mobility data perfectly fit with the data stream concept and share the following characteristics and processing requirements [11, 28, 56]:

- A data stream is an ordered, potentially unbounded sequence of data items called *stream elements*.
- Stream elements are generated by an active external source (mobile object).
- Spatio-temporal data streams have *explicit* ordering—i.e., stream elements provide timestamp indicating their generation by an active external source.
- Stream elements are pushed by an active external source and arrive continuously in the system.
- Stream elements are accessed *sequentially*—the stream element that has already arrived and has been processed cannot be retrieved without being explicitly stored.
- Query over data streams runs continuously and returns new results as a new stream element arrives.
- Real-time and batch processing in a single paradigm.

To efficiently support processing of big spatio-temporal data streams in a cluster computing, a streaming computation framework that supports spatio-temporal data types and operations, and scales transparently to large clusters, is required.

Consequently, our goal is to achieve an integrated query processing approach that runs SQL-like expressions continuously and incrementally over data before that data is stored in the persistent storage. The key concept here is that mobility streaming data and persistent data are not intrinsically different—the persistent data are simply streaming data that have been entered into the persistent structures. In other words, query processing could be exclusively preformed on persistent data, exclusively on streams, or on a combination of streams and persistent data [25]. In the first case, a query produces a relation as the output and has the exact same semantics as in SQL. We refer to queries over relations as relational *snapshot* queries because they operate on a snapshot of the relations at a given time—they produce an answer and they terminate. In other cases however, a query produces a stream as the output. Since a stream is unbounded, a query that produces a stream never ends and is therefore called a *continuous* query.

Therefore, there is obviously a clear need for highly scalable spatio-temporal stream computing framework that can operate at high data rates and process massive amounts of big mobility data.

3.2 MobyDick—A Prototype of Distributed Framework for Big Mobility Data Processing and Analytics

The framework we propose is a cornerstone towards efficient real-time managing and monitoring of mobile objects through distributed spatio-temporal streams processing on large clusters. A prototype[1] implementation is rooted in a new stream processing model that overcomes the challenges of current distributed stream processing models and enables seamless integration with batch and interactive processing.

3.2.1 Data Model

Data model provides a formalism consisting of a notation for describing data of interest, set of operations for manipulating these data and a set of predicates. Spatio-temporal data model is a data model representing the temporal evolution of spatial objects over time [55]. If these evolutions are continuous, one speaks about mobile objects and represents them by spatio-temporal data types like mobile point (for example, recording the route of a car), mobile points (for example, representing movements of the fleet of trucks), mobile line (for example, representing movement

---

[1] Available on https://bitbucket.org/DarioOsm/mobydick—it relies on operations and functions of Apache Flink DataStream library [6] and JTS Topology Suite [43].

Table 3.1 Type system—abstract model

Type constructor	Signature	Target sort
integer, real *string, boolean* *point, multipoint,* *linestring,*		→ BASE
polygon		→ SPATIAL
instant, interval, *period*		→ time
intime	BASE ∪ SPATIAL	→ TEMPORAL
mobile	BASE	→ TEMPORAL
mobile	SPATIAL	→ TEMPORAL

of the train), etc. Spatio-temporal data types enable the user to describe the dynamic behavior of spatial and transportation objects over time.

Mobile object is a spatio-temporal object that continuously changes its location and/or shape. Depending on the particular mobile objects and applications, the movements of mobile objects may be subject to constraints. We distinguish between two movement patterns: *unconstrained* movement in Euclidean space and *constrained* movement within spatially embedded networks. In what follows, we focus on unconstrained, free movement in Euclidean space.

Due to the facts that our prototype should explicitly support processing of big mobility data, we will slightly redefine abstract model previously defined in Chap. 2, Sect. 2.3.1 and specify discrete model (Table 3.2).

All static data types are made uniformly time dependent by introducing a type constructor ***mobile*** (Table 3.1). For a given static data type α, *mobile* returns the type whose values are partial functions from the time domain $\mathbb{T}$ into α. The domain for type *mobile* (α) is

$$\mathscr{D}_{mobile(\alpha)} := \{f \mid f : \mathscr{D}_{instant} \nrightarrow \mathscr{D}_\alpha\} \tag{3.1}$$

The signatures of all temporally lifted operations, i.e. their argument sorts and the target sorts should be refined accordingly:

$$\begin{aligned}
\textbf{\textit{inside}} : mobile(point) \times polygon &\rightarrow mobile(boolean) \\
\textbf{\textit{inside}} : point \times mobile(polygon) &\rightarrow mobile(boolean) \\
\textbf{\textit{inside}} : mobile(point) \times mobile(polygon) &\rightarrow mobile(boolean)
\end{aligned}$$

If we abbreviate the formally defined notation $mobile(\alpha)$ with $m\alpha$, a lifted operation ***inside*** has the following signatures:

Table 3.2 Type system—discrete model

Type constructor	Signature	Target sort
integer, real string, boolean point, multipoint, linestring,		→ BASE
polygon		→ SPATIAL
instant, interval, period		→ TIME
intime	BASE ∪ SPATIAL	→ TEMPORAL
sttuple	((BASE⁺ ∪ SPATIAL ∪ TIME)×TIME)	→ STTUPLE
ststream	STTUPLE	→ STSTREAM
now, unbounded, range		→ WINDOW
temporal	BASE	→ TEMPORAL
temporal	STSTREAM × WINDOW	→ TEMPORAL

inside : *mpoint × polygon*   → *mboolean*
inside : *point × mpolygon*   → *mboolean*
inside : *mpoint × mpolygon*   → *mboolean*

The discrete type system replaces **mobile** constructor with **temporal** (Table 3.2).

The abstract spatial data types framework [34, 36] together with [33] has influenced precise discrete specification (Fig. 3.1).

For data type *instant* we explore TM_Instant as specified in [33], and it could be directly implemented with Java/Scala Timestamp type. The type TimeInterval represents *interval* and inherits from the type TM_Period [33]. As such it implements all attributes, operations and associations inherited from that type and is extended with two attributes representing whether time period is left or right closed. The type *period* is represented using type TM_Period, with two attributes representing begin and end of period, both of type TM_Instant.

We model *intime* by temporal data type IntimeObject composed of position in time and space, thus representing pair (*instant, value*).

Result of many operations on spatio-temporal (mobile) objects is type of TemporalReal, whose units are composed of time interval and quadratic function represented with an array of its coefficients. Operations localMin and localMax for type RealUnit return values of local minimum and local maximum for a quadratic function valid on that unit, while operations min and max return overall minimum and maximum values for object of this type. Operations valueAtTime and timeAtValue return real value at a given time instant and an array of time instants that lists in ascending order the times at which given real value is reached, respectively.

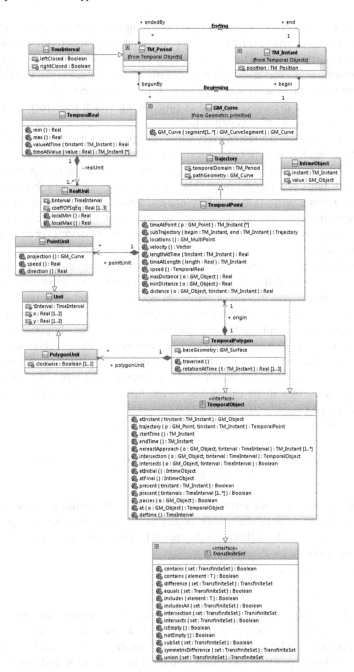

Fig. 3.1 Discrete model—UML class diagram

Our concept of trajectory implies projection of temporal point onto 2D plane which is valid only for a period of time when temporal point is defined. It is a spatial curve representing the projection of the trajectory over time in which relationship to time or path orientation is not preserved and repeated traversal of the same place at different times is not reflected. Besides that, it contains attribute temporalDomain, of type TimePeriod, that presents a period of time during which trajectory, as well as temporal point consequently, is defined.

As we have mentioned already, every temporal type is presented in a form of sliced representation built from unit types consisting of a time interval and a simple linear function. This is modeled with type UnitType where arrays x and y represent linear function. For units of temporal point this is linear change of x and y coordinates, and we can perform projection onto 2D plane and determine speed and direction that hold during unit's time interval. This is implemented with type PointUnit which inherits from the type UnitType and also implements operations projection, speed and direction.

All spatio-temporal objects share a set of common operations. For this reason all spatio-temporal data types shall implement common operations, as well operations that are specific for a particular spatio-temporal data type. Similar behavior is expressed through the interface[2] TemporalObject which specifies common operations for all spatio-temporal data types. Number of operations, specified by TemporalObject, are precisely defined in [26, 27, 34]—they have the same signature and meaning in this type system as well. For this interface and TemporalPoint and TemporalPolygon temporal types, we will briefly describe just those operations with different signature or meaning and the new operations that we propose.

The operation

```
nearestApproach(o:GM_Object, tinterval:TimeInterval
                ):TM_Instant[1..*]
```

shall return an array of time instants that lists in ascending order the time or times of the nearest approach of the mobile object to a given static spatial object. The parameter tinterval shall restrict the search to a particular time interval.

Two operations

```
startTime(): TM_Instant
endTime(): TM_Instant
```

return the result of type TM_Instant which represents start and end time value of the period in which the type TemporalPoint is defined, respectively.

The operations atInitial and atFinal return the type IntimeObject with two attributes: position and shape of mobile object in time and space, which

---

[2]Scala programming language does not support interfaces, but enables multiple inheritance by implementing interfaces as *traits*. For this reason TemporalObject is implemented as a trait that enables TemporalPoint to inherit from two types.

is defined with `startTime` operation for `initial` and `endTime` operation for `atFinal`.

The `TemporalPoint` is composed of `PointUnit` set, whereas operations `timeAtPoint`, `subTrajectory` and `velocity` have the same meaning and signature as corresponding operations in [35]. Operation `locations` returns isolated points at the beginning and end of each unit's time interval. We introduce operation `distance` exclusively as a measure of linear space between two spatial objects, and operation `length` as a measure of trajectory's total length. For this reason we introduce the following operations on `TemporalPoint`:

```
lengthAtTime(tinstant:TM_Instant):Real
timeAtLength(length:Real):TM_Instant
maxDistance(o:GM_Object):Real
minDistance(o:GM_Object):Real
distance(o:GM_Object, tinstant:TM_Instant):Real
```

Operation `lengthAtTime` returns total trajectory's length from the beginning of the trajectory until the given time instant `tinstant` and `timeAtLength` returns the time at which the trajectory's total length from the beginning reaches the given value. Operations `minDistance` and `maxDistance` returns the distance of the nearest and furthest approach of the temporal point to a given spatial object. Operation `distance` returns the distance of the temporal point from a given spatial object at a given time instant `tinstant`.

As we have already pointed out, none of Big Data streaming frameworks and platforms has built-in spatio-temporal capabilities. However, we choose Apache Flink as an underlying platform due to its efficient, parallel fault recovery mechanism, and a unique ability to combine nearly real-time and batch in-memory processing in a unified programming framework, which is common in IFP applications involving both real-time and historical mobility data. Another important reason is the fact that we have already pointed out (Chap. 2, Sect. 2.1): spatio-temporal streams are *explicitly* timestamped. As an important consequence, analyzing spatio-temporal events with respect to their *event* time is by far the most interesting, compared to analyzing them with respect to the time when they arrive in the system. Therefore, underlying distributed streaming engine should be configurable towards processing mobility data based on event time.

The reason is that event time for a given tuple is immutable, but processing time changes constantly for each tuple as it flows through the pipeline and time advances ever forward. During processing, the realities of the systems in use result in an inherent and dynamically changing amount of skew between the two temporal domains. Global progress metrics, such as punctuation or watermark,[3] provide a nice way to visualize temporal domain skew (Fig. 3.2). This dynamic variance in skew is

---

[3]Watermark is a special event generated at stream sources that coarsely advances event time. A watermark for time instant τ states that event time has progressed to τ in that particular stream, meaning that no events with a time instant smaller than τ can arrive any more.

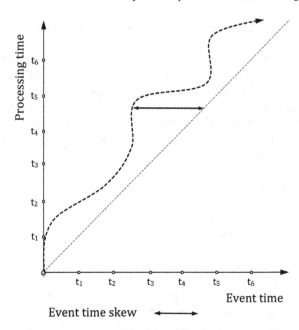

Fig. 3.2 Time domain skew

very common in distributed data processing systems, and plays a big part in defining what functionality is necessary for providing correct processing and results [1].

3.2.2 Apache Flink

Apache Flink [6] is a platform for efficient, distributed, general-purpose data processing with powerful programming abstractions, a high-performance run-time, and automatic program optimization. It is a community-developed fork of Stratosphere [2]—a cluster computing platform aiming to bridge the gap between MapReduce-like systems and shared-nothing parallel DBMS. By extending and generalizing MapReduce framework it goes far beyond batch applications and enables to combine multiple types of computation (batch, streaming, SQL-like expressions, machine learning, etc.) that might previously have required different Big Data engines.

The core of Flink is a distributed streaming dataflow engine that provides data distribution, communication, and fault tolerance for distributed computations over data streams and executes programs in a data-parallel and pipelined manner. Programs are automatically compiled and optimized into dataflow programs that are executed in a *computing cluster* or cloud environment. It has advanced the state of the art in data management in several aspects:

(i) A data programming model based on second-order functions to abstract paral-
lelization.
(ii) A method that uses static code analysis of user-defined functions to achieve
goals similar to database query optimization in a UDF-heavy environment.
(iii) Abstractions to integrate iterative processing in a dataflow system with good
performance.
(iv) An extensible query language and underlying operator model.

Flink is a layered system (Fig. 3.3)—the different layers of the stack build on
top of each other and raise the abstraction level of the program representations they
accept [6]:

- The API layer implements multiple APIs that generate JobGraphs (DAGs) for their
programs through separate compilation processes.
- The optimizer and common API layer takes programs in the form of operator
DAGs. The DataSet API uses an optimizer to determine the optimal plan for the
program, while the DataStream API uses a stream builder. The all operators (e.g.,
Map, Reduce, Filter, Join, etc.) are data type agnostic. The concrete types and their
interaction with the run-time is specified by the higher layers.
- The run-time layer receives a program in the form of a JobGraph—a generic
parallel data flow with arbitrary tasks that consume and produce data streams. The
JobGraph is executed according to a variety of available deployment options (i.e.,
local, remote, YARN, etc.).

Flink natively supports stateful computation, data-driven windowing semantics
and iterative stream processing. However, similarly to all existing, general-purpose,
distributed stream processing engines [8, 10, 12, 17, 47, 53], etc., it also possesses a
crucial limitation for real-time processing of big spatio-temporal mobility data: The
DataStream is the basic data abstraction and represents a continuous, parallel,
immutable stream of data of a certain type. It is a statically typed object parametrized
by an element type, which is itself native Java/Scala data type. Consequently, the
support for spatio-temporal data types and related operations is completely missing.

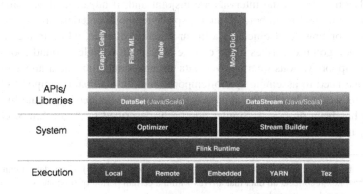

Fig. 3.3 The Apache Flink stack with MobyDick library

Nevertheless, Flink has a number of unique features when compared to other available distributed stream processing engines, making it suitable as an underlying platform for spatio-temporal data streams:

(i) High performance and low latency—its data streaming run-time achieves high throughput rates and low latency.

(ii) Processing data streams as true streams, i.e., data elements are immediately "pipelined" though a streaming program as soon as they arrive.[4]

(iii) Three different notions of time: *processing* time, *event* time, and *ingestion* time. Event time decouples the program semantics from the actual serving speed of the source and the processing performance of the system. Event time windows compute correct results, even if events arrive out-of-order of their timestamp which is common if a data stream gathers events from distributed sources.

(iv) Streaming and batch processing in one system—one common run-time is used both for data streaming applications and batch processing applications.[5] Flink has a unique ability to combine real-time and batch in-memory processing in a unified programming framework, which is common in IFP applications involving both real-time spatio-temporal and persistent spatial data.

(v) Fault-tolerance mechanism via lightweight distributed snapshots [18], allowing the system to maintain high throughput rates and provide strong consistency guarantees.

(vi) Exactly-once semantics for stateful computations—checkpointing mechanism ensures *exactly-once* semantics for the state in the presence of failures.

(vii) Iterations and delta iterations—dedicated support for iterative computations and faster convergence through delta iterations by exploring computational dependencies.

3.2.3 Spatio-Temporal Queries

Implemented discrete model presents a spatio-temporal object as a set of spatio-temporal units whose time intervals are disjoint and, if adjacent, their values are different. In the case of temporal point, the spatio-temporal unit describes movement of a corresponding point object as a linear function of time, while in the case of temporal polygon it describes linear change of a rotation angles around coordinate axes. To support various queries over data streams containing data about mobile objects, we need to assemble spatio-temporal units into a spatio-temporal object. In previous works [26, 27] the approach based on user defined aggregate function (UDAF) has been proposed. Similar approach, described in Algorithm 3, can be elegantly implemented in the form of SQL-like expressions.

---

[4]On the contrary, some cluster computing frameworks (Spark, Trill, etc.) wraps data streams into mini-batches, i.e., it collects all data that arrives within a certain period of time and runs a regular batch program on the collected data.

[5]Batch processing applications run efficiently as special cases of stream processing applications.

Algorithm 3 Spatio-temporal query

Input: $\mathscr{S}_{\sigma\theta}$,   ▷spatio-temporal stream
     $\mathscr{W}$ ▷time-based window
$k\mathscr{S}_{\sigma\theta} \leftarrow \kappa_{key}\mathscr{S}_{\sigma\theta}$ ▷logically partitioned *keyed* stream
$w\mathscr{S}_{\sigma\theta} \leftarrow \psi k\mathscr{S}_{\sigma\theta}$ ▷windowed stream
$m\mathscr{S} \leftarrow \alpha_{mobile} w\mathscr{S}_{\sigma\theta}$ ▷stream of mobile objects
$m\mathscr{S} \leftarrow \phi_{f:B^k\rightarrow B} m\mathscr{S}$ ▷retain mobile objects for which function f returns *true*
return $\mu_{f:X\rightarrow Y} m\mathscr{S}$ ▷equivalent to SELECT clause

---

For validation purposes we used GeoLife[6]—a well known real-life mobility data collected by Microsoft Research Asia [61, 62]. A trajectory of this dataset is represented by a sequence of time-stamped points, each of which contains the information of latitude, longitude and altitude.

Spatio-temporal `DataStream` in Flink is a possibly unbounded immutable collection of data items of the same type, and before those steps we have to specify a data source of a stream with spatio-temporal data stream schema $\Sigma_{\sigma\theta}$ and temporal ordering f_Ω according to *event* time, as previously defined in Chap. 2, Sect. 2.1.

Therefore, we define a Scala `case` class according to the spatio-temporal data stream schema $\Sigma_{\sigma\theta}$ and use it in stream execution environment for adding a data source or simply for mapping a data read from a socket to a derived data stream:

```
case class GeoLife(id:Int,
              location:Point,
              timestamp:Timestamp)
```

In this section we perform an evaluation and validation of our approach by formulating a number of queries that had been specified in natural language before.

Spatio-temporal `DataStream` can be created either from data sources (file-based, collection-based, Apache Kafka, RabbitMQ, Twitter Streaming, etc.) or by applying high-level operations on other `DataStream`. In our experimental setup we simulate a real-world scenario by using Apache Kafka connector to access GeoLife mobility data as spatio-temporal data streams:

```
val env =
StreamExecutionEnvironment.getExecutionEnvironment
val stream = env.addSource (new KafkaSource ...)
```

In order to work with event time semantics, it is necessary to set execution stream environment time characteristic to an event time:

```
env.setStreamTimeCharacteristic(TimeCharacteristic.EventTime)
```

We also define how timestamps relate to events (e.g., which tuple field is the event timestamp). According to temporal ordering f_Ω of spatio-temporal data streams on *event* time, we apply the `assignAscendingTimestamps` function:

---

[6]This GPS dataset was collected by 182 users in a period of over five years, from April 2007 to August 2012. It contains approximately 20 million points with a total distance of about 1.2 million kilometers and a total duration of 48,000+ h. The data were logged in over 30 cities in China, USA, and Europe.

```
val geoLifeStream : DataStream[GeoLife] =
    stream.map{tuple => GeoLife(tuple)}
geoLifeStream.assignAscendingTimestamps(
                tuple => tuple.timestamp.getTime)
```

Q_1: *mobileContinuously report mobile objects (id and position) within the area of interest.*

```
val coords = : Array[Coordinate] = Array(...)
val geomFactory = new GeometryFactory()
val ring = geomFactory.createLinearRing(coords)
val holes : Array[LinearRing] = null
vlr areaOfInterest = geomFactory.createPolygon(ring, holes)
```

```
val q1 = geoLifeStream
    .filter(geolife => geolife
                .position
                .within(areaOfInterest))
      .map(geolife => (geolife.id, geolife.position))
      q1.print
```

Q_2: *Continuously each minute, report location of mobile objects which have traveled more than 3 Km in past 10 minutes.*

```
val q2 = geoLifeStream
        .keyBy(0)
        .timeWindow(Time.of(10, TimeUnit.MINUTES),
                Time.of(1, TimeUnit.MINUTES))
        .apply { MobyDick.mobilePoint _ }
        .filter(mo => mo
                    .location
                    .lengthAtTime(mo.location.endTime) > 3000
            )
        .map(mo => mo.location.atFinal.geom)
      q2.print
```

Q_3: *For each mobile object, find its minimal distance from the point of interest during last half an hour.*

```
val easting = ...
val northing = ...
val point = new Coordinate(easting, northing)
val factory = new GeometryFactory()
val pointOfInterest = factory.createPoint(point)
```

```
val q3 = geoLifeStream
      .keyBy(0)
      .timeWindow(Time.of(30, TimeUnit.MINUTES))
      .apply { MobyDick.mobilePoint _ }
      .map(mo => (mo.id,
              mo.position
                .minDistance(pointOfInterest)
              )
          )
      q3.print
```

Q₄: *Find all mobile objects (id, position and distance traveled) that have traveled more than 10 Km during the last hour.*

```
val  q4 = geoLifeStream
      .keyBy(0)
      .timeWindow(Time.of(60, TimeUnit.MINUTES))
      .apply { MobyDick.mobilePoint _ }
      .filter(mo => mo
                  .location
                  .lengthAtTime(
                   mo.location.endTime)
                  > 10000
             )
      .map(mo => (mo.id,
              mo
              .location
              .lengthAtTime(mo.location.endTime)
              )
          )
        q4.print
```

Q₅: *Find trajectories of the mobile objects that have been less than 500 m from a point of interest within last 15 min.*

```
val  q5 = geoLifeStream
      .keyBy (0)
      .timeWindow(Time.of(15, TimeUnit.MINUTES))
      .apply { MobyDick.mobilePoint _ }
      .filter (mo => mo
                   .location
                   .distance(
                    pointOfInterest,
                    mo.location.endTime)
                   .asInstanceOf[Double] < 500
              )
      .map (mo => (mo.id,
               mo
               .location
               .subTrajectory(
                mo.location.startTime,
                mo.location.endTime)
               )
           )
         q5.print
```

3.3 Related Work

There is plenty of work in the areas of mobility analysis, acquiring mobile traffic data, mapping it to road-networks, using the data for traffic prediction, etc. However, the status of related work is rather limited, with just a few works barely similar to ours in some particular respects. To get a broader overview, we classify related work into three main categories: (i) distributed spatial and spatio-temporal batch systems, (ii) centralized DSMS-based systems, and (iii) distributed DSMS-based systems.

3.3.1 Distributed Spatial and Spatio-Temporal Batch Systems

SpatialHadoop prototype [23] is an extension to Hadoop that pushes spatial data inside Hadoop core functionality by adopting a layered design. The storage layer employs a two-level index structure: global spatial index partitions data across computation nodes, while the local spatial index organizes data inside each node. The operations layer encapsulates the implementation of spatial operations that take advantage of the spatial indexes and the new components in the MapReduce layer. SpatialHadoop is initially equipped with a rather limited set of spatial operations: range queries, k-nearest-neighbor queries, and spatial join, and does not include any spatio-temporal data types and operations for mobility data.

Sphinx [22] is a distributed system which uses a standard SQL interface to process big spatial data. Sphinx extends the core of Cloudera Impala [39] to provide real-time SQL query processing on spatial data by adding spatial data types, indexes and query processing for efficient processing of spatial data.

Papers [54] and [59] envision GeoSpark—in-memory cluster computing framework for processing large-scale spatial data by providing native support for spatial data indexing and query processing algorithms in Apache SparkApache!Spark. GeoSpark introduces two new layers: Spatial Layer and Spatial Query Processing Layer. Spatial Layer consists of three novel Spatial Resilient Distributed Datasets (SRDDs) which extend regular Spark RDDs to support spatial objects. The Spatial Query Processing Layer efficiently executes spatial query processing algorithms (e.g., Spatial Range, Join, KNN query) on SRDDs.

Magellan [29] is an open source library that uses Apache Spark as the underlying execution engine. Magellan facilitates geospatial queries towards solving problems of dealing with geospatial data at scale.

GeoMesa [19] is an open-source, distributed, spatio-temporal database built on top of the Apache Accumulo—distributed key/value store based on the BigTable technology from Google. It implements a custom Geohash algorithm to combine spatio-temporal domain into a single-dimension lexicographic space provided by Accumulo [24].

Parallel Secondo [40] is a parallel spatial DBMS which uses Hadoop Apache!Hadoop as a distributed task scheduler, while all storage and query processing are done by spatial DBMS engines running on cluster nodes. It proposes parallel data types and their relevant operators to make end-users represent parallel queries like common queries, as well as a simple and independent distributed file system to transfer data among database engines directly. To remove as much as possible unnecessary transform and transfer, intermediate data are directly transferred among single databases via their own parallel file system, without passing through HDFS.

Unlike the Hadoop framework composed by nodes, Parallel Secondo is deployed as data servers.

Distributed Secondo [49] uses Apache Cassandra for data storage, and Secondo as a query processing engine. It is a distributed system consisting of three different node types: storage nodes, query processing nodes, and management nodes. Storage nodes are running Apache Cassandra to provide a highly available data storage. Query processing nodes run Secondo and Query executor for the distribution of the queries, whereas management nodes also run Secondo for importing and exporting data and to specify the queries that should be executed.

The work presented in [37] proposes MOIST, whose baseline is a recursive spatial partitioning indexer built upon BigTable. To reduce update and query contention, MOIST groups nearby objects of similar trajectory into the same school, and keeps track of only the history of school leaders. This dynamic clustering scheme can eliminate redundant updates and hence reduce update latency. To improve history query processing, MOIST keeps some history data in memory, while it flushes aged data onto parallel disks in a locality-preserving way.

Both PRADASE [41] and CloST [57] are Hadoop-based storage systems for big spatio-temporal data analytics, based on MapReduce framework. CloST is targeted at fast data loading, scalable spatio-temporal range query processing and efficient storage usage to handle very large historical spatio-temporal datasets. It includes a simple data model which has special treatments on three core attributes including an object id, a location and a time. Based on this data model, CloST hierarchically partitions data using all core attributes which enables efficient parallel processing of two simple types of spatio-temporal queries: single-object queries and all-object queries.

However, all the above mentioned systems follow *batch-oriented* processing paradigm: spatial (or spatio-temporal in [40, 49]) data are persistently stored and indexed on disk before being processed. Thus, they rely on outbound, pull-based paradigm for processing of large-scale *historical* spatial data. Consequently, we consider them orthogonal to our approach and research; we rely on inbound, push-based processing paradigm and provide a foundation for real-time processing of large-scale, spatio-temporal mobility data.

3.3.2 *Centralized DSMS-Based Systems*

The PLACE server [46] views spatial data streams as automatically changing relations incrementally calculating results and producing positive and negative updates of the result. It supports continuous query processing of spatio-temporal streams and allows the user to construct complex queries out of simpler operators such as inside or kNN (k-nearest-neighbor). The PLACE continuous query processor extends the processing of continuous sliding window queries to accommodate for *predicate-based* window queries.

In [31, 60] new data types and operations to support geostreaming applications have been proposed, together with their semantics and embedding into a database language such as SQL. The proposed system is compatible with a data type based spatio-temporal database system. It can be used to support continuous real-time queries on evolving spatio-temporal extents through the combination of these new data types and operations with SQL. This work is one of the first that formally and successfully merges geostreams with moving objects, towards a full-fledged geostream management system.

The work presented in [4] utilizes a generic approach to extend a streaming system for a particular application domain. It is focused on integrating user-defined modules (UDM) within the query pipeline of a DSMS. This approach relies on a standard spatial library and supports a native approach to deal with spatial attributes as first-class citizens, to reason about the spatial properties of incoming events and to provide consistency guarantees over space and time.

A novel formal framework and an implementation approach to the management of spatio-temporal objects considering them as geospatial data streams is proposed in [27]. The work is bridging the formal frameworks of geospatial databases, spatio-temporal databases and data streams, at the abstract level. The proposed formal framework is implemented as a prototype in a pure DSMS environment.

A real-time evaluation of vehicle traces over road networks and their collective representation as traffic data streams is advocated in [51, 52]. The work considers a geostreaming approach to road traffic data, enhanced with methods for extracting privacy-preserving driving patterns and evolving phenomena network-wide. The proposed framework is mostly geared towards a multi-resolution representation of traffic streams and props up multi-modal diffusion of custom, succinct analytics to potential users.

The approaches taken in these works share a common feature: they consider spatio-temporal data streams management and processing within *centralized* architecture, nor within distributed computing nor Big Data framework.

3.3.3   Distributed DSMS-Based Systems

The work presented in [32] focuses on predicting drivers travel times in a large urban area from sparse GPS traces, collected in real-time from vehicles, and over a very large network. An on-line expectation maximization algorithm is representative of a large class of Machine Learning algorithms. This algorithm is the core of an estimation pipeline deployed inside the Mobile Millennium traffic information system [15], which gathers GPS observations from participating vehicles and produces estimates of the travel times on the road network. However, it uses Spark Streaming [10] in *as-is* manner, exclusively for a particular machine learning algorithm, without specific spatio-temporal data types and operation. The work does not exploit spatio-temporal nature of mobility data we exploit to make Apache Flink "mobility-aware".

A pilot traffic-monitoring system prototype [14] is based on a component-based distributed stream processing platform [13] that generates dynamic, multi-faceted views of transportation information, using real vehicle GPS and road network data. The system continuously derives current traffic statistics, and provides useful transportation information such as average speed, shortest-time routes from real-time observed and inferred traffic conditions. Similarly to the previous work, distributed stream processing platform is explored in *as-is* manner—the spatio-temporal nature of mobility data and optimization for spatio-temporal processing has been neglected.

ElaStream [38] is a general cloud-based streaming infrastructure that enables huge parallelism by means of the divide, conquer and combine paradigm, which resembles the two layer structure of Map and Reduce. ElaStream framework is dedicated to streaming and is different from MapReduce: (i) it does not depend on persistent storage, (ii) it is based on a push (rather than pull) model, and (iii), it envisions the components of a streaming application (e.g. dealing with streaming operators). The authors argue that the large-scale spatio-temporal streaming applications can be best served on cloud compared to any other infrastructure. An overall design of ElaStream is based on the premise that cloud is the most appropriate infrastructure in which is possible to support stream data processing. However, extensive research is required to adapt such applications to obtain all the automation, flexibility, parallelism, and cost efficiency that cloud offers.

Tornado [42] is a distributed in-memory spatio-textual stream processing system that extends Apache Storm [8]. To efficiently process spatio-textual streams, Tornado introduces an adaptive spatio-textual indexing layer to dynamically re-distribute the processing across the system according to changes in the data distribution and/or query workload. Query language follows the philosophy of SQL and relational algebra by offering simple declarative spatial, textual, and semantic building block operators and predicates that are composable to form complex spatio-textual queries. However, Tornado is mainly concerned with a rather limited set of spatio-textual queries: range and kNN select and join predicates, limited history, and temporal predicates.

Our proposed framework is generic—it is envisioned and designed as a research vehicle towards supporting a wide range of IFP applications involving large-scale mobility data, and relies upon a comprehensive set of spatio-temporal data types and related operations. The approach presented in this paper also differs in that we focus on a rich set of spatio-temporal data types and operations towards mobility data management, where necessary batch processing of historical spatial and spatio-temporal mobility data runs efficiently as a special case of stream processing.

3.4 Final Remarks

In this chapter, we presented the design and implementation of a novel, in-memory parallel and distributed system that supports real-time processing and analysis of big mobility data derived from spatio-temporal data streams. The main contribution is a precise and comprehensive formal framework for mobile objects within data stream and Big Data paradigms. The prototype is built on top of a distributed streaming dataflow engine and extends it with a type system consisting of the set of spatio-temporal data types and corresponding operations. The system provides native support for spatio-temporal query processing to efficiently analyze big mobility data in real-time.

The proposed framework is generic—it could be used for the extension of other available stream processing engines.

We illustrated and validated our proposed formal framework by different, representative spatio-temporal queries on big mobility data streams. We also showed that many operations on big mobility data can be realized efficiently using our spatio-temporal built-in components.

However, there are several important issues regarding this topic which need to be addressed in the near future. We will in particular focus on the following conceptual and implementation issues: (i) optimization strategies including spatio-temporal indexing layer, (ii) spatio-temporal predicates, and (iii) distributed predicate windows.

References

1. Akidau, T., Bradshaw, R., Chambers, C., Chernyak, S., Fernández-Moctezuma, R., Lax, R., McVeety, S., Mills, D., Perry, F., Schmidt, E., Whittle, S.: The dataflow model: a practical approach to balancing correctness, latency, and cost in massive-scale, unbounded, out-of-order data processing. PVLDB **8**(12), 1792–1803 (2015). http://www.vldb.org/pvldb/vol8/p1792-Akidau.pdf
2. Alexandrov, A., Bergmann, R., Ewen, S., Freytag, J., Hueske, F., Heise, A., Kao, O., Leich, M., Leser, U., Markl, V., Naumann, F., Peters, M., Rheinländer, A., Sax, M.J., Schelter, S., Höger, M., Tzoumas, K., Warneke, D.: The Stratosphere platform for big data analytics. VLDB J. **23**(6), 939–964 (2014). http://dx.doi.org/10.1007/s00778-014-0357-y
3. Ali, M.H., Gerea, C., Raman, B.S., Sezgin, B., Tarnavski, T., Verona, T., Wang, P., Zabback, P., Kirilov, A., Ananthanarayan, A., Lu, M., Raizman, A., Krishnan, R., Schindlauer, R., Grabs, T., Bjeletich, S., Chandramouli, B., Goldstein, J., Bhat, S., Li, Y., Nicola, V.D., Wang, X., Maier, D., Santos, I., Nano, O., Grell, S.: Microsoft CEP server and online behavioral targeting. PVLDB **2**(2), 1558–1561 (2009)
4. Ali, M.H., Chandramouli, B., Raman, B.S., Katibah, E.: Spatio-temporal stream processing in microsoft streaminsight. IEEE Data Eng. Bull. **33**(2), 69–74 (2010)
5. Apache Software Foundation: Spark. http://spark.apache.org
6. Apache Software Foundation: Apache Flink (2016). http://flink.apache.org/
7. Apache Software Foundation: Apache Samza (2016). http://samza.apache.org
8. Apache Software Foundation: Apache Storm (2016). http://storm.apache.org
9. Apache Software Foundation: S4 (2016). http://incubator.apache.org/s4
10. Apache Software Foundation: Spark Streaming (2016). http://spark.apache.org/streaming
11. Babcock, B., Babu, S., Datar, M., Motwani, R., Widom, J.: Models and issues in data stream systems. In: Popa, L., Abiteboul, S., Kolaitis, P.G. (eds.) PODS. pp. 1–16. ACM (2002)
12. Balazinska, M., Balakrishnan, H., Madden, S., Stonebraker, M.: Fault-tolerance in the Borealis distributed stream processing system. ACM Trans. Database Syst. **33**(1), 3:1–3:44 (2008)

13. Ballard, C., Brandt, O., Devaraju, B., Farrell, D., Foster, K., Howard, C., Nicholls, P., Pasricha, A., Rea, R., Schulz, N., Shimada, T., Thorson, J., Tucker, S., Uleman, R.: IBM InfoSphere Streams: Accelerating Deployments with Analytic Accelerators. IBM (2014)
14. Biem, A., Bouillet, E., Feng, H., Ranganathan, A., Riabov, A., Verscheure, O., Koutsopoulos, H.N., Moran, C.: IBM InfoSphere Streams for scalable, real-time, intelligent transportation services. In: Elmagarmid, A.K., Agrawal, D. (eds.) Proceedings of the ACM SIGMOD International Conference on Management of Data, SIGMOD 2010, Indianapolis, Indiana, USA, June 6–10, 2010. pp. 1093–1104. ACM (2010). http://doi.acm.org/10.1145/1807167.1807291
15. California Center for Innovative Transportation: The Mobile Millennium Project (2016). http://traffic.berkeley.edu
16. Capriolo, E., Wampler, D., Rutherglen, J.: Programming Hive, 1st edn. O'Reilly Media, Inc., California (2012)
17. Chandramouli, B., Goldstein, J., Barnett, M., DeLine, R., Platt, J.C., Terwilliger, J.F., Wernsing, J.: Trill: a high-performance incremental query processor for diverse analytics. PVLDB 8(4), 401–412 (2014). http://www.vldb.org/pvldb/vol8/p401-chandramouli.pdf
18. Chandy, K.M., Lamport, L.: Distributed snapshots: determining global states of distributed systems. ACM Trans. Comput. Syst. 3(1), 63–75 (1985). http://doi.acm.org/10.1145/214451.214456
19. Commonwealth Computer Research, Inc.: GeoMesa (2016). http://www.geomesa.org
20. Condie, T., Conway, N., Alvaro, P., Hellerstein, J.M., Elmeleegy, K., Sears, R.: Mapreduce online. In: Proceedings of the 7th USENIX Symposium on Networked Systems Design and Implementation, NSDI 2010, Apr 28–30, 2010, San Jose, CA, USA. pp. 313–328. USENIX Association (2010)
21. Dean, J., Ghemawat, S.: MapReduce: Simplified data processing on large clusters. In: Brewer, E.A., Chen, P. (eds.) OSDI. pp. 137–150. USENIX Association (2004)
22. Eldawy, A., Elganainy, M., Bakeer, A., Abdelmotaleb, A., Mokbel, M.: Sphinx: Distributed execution of interactive sql queries on big spatial data. In: Proceedings of the 23rd SIGSPATIAL International Conference on Advances in Geographic Information Systems. pp. 78:1–78:4. GIS '15, ACM, New York, NY, USA (2015). http://doi.acm.org/10.1145/2820783.2820869
23. Eldawy, A., Mokbel, M.F.: A demonstration of SpatialHadoop: an efficient MapReduce framework for spatial data. PVLDB 6(12), 1230–1233 (2013)
24. Fox, A., Eichelberger, C., Hughes, J., Lyon, S.: Spatio-temporal indexing in non-relational distributed databases. In: Hu et al. [30], pp. 291–299. http://dx.doi.org/10.1109/BigData.2013.6691586
25. Franklin, M.J., Krishnamurthy, S., Conway, N., Li, A., Russakovsky, A., Thombre, N.: Continuous analytics: Rethinking query processing in a network-effect world. In: CIDR (2009). www.crdrdb.org
26. Galić, Z., Mešković, E., Križanović, K., Baranović, M.: OCEANUS: a spatio-temporal data stream system prototype. In: Proceedings of the Third ACM SIGSPATIAL International Workshop on GeoStreaming. pp. 109–115. IWGS '12, ACM, New York, NY, USA (2012). http://doi.acm.org/10.1145/2442968.2442982
27. Galić, Z., Baranović, M., Križanović, K., Mešković, E.: Geospatial data streams: formal framework and implementation. Data Knowl Eng 91, 1–16 (2014)
28. Golab, L., Özsu, M.T.: Data Stream Management. Synthesis lectures on data management. Morgan Claypool Publishers, San Rafael, CA (2010)
29. Hortonworks: Magellan: Geospatial Analytics on Spark. http://hortonworks.com/blog/magellan-geospatial-analytics-in-spark (2016)
30. Hu, X., Lin, T.Y., Raghavan, V.V., Wah, B.W., Baeza-Yates, R.A., Fox, G., Shahabi, C., Smith, M., Yang, Q., Ghani, R., Fan, W., Lempel, R., Nambiar, R. (eds.) In: Proceedings of the 2013 IEEE International Conference on Big Data, 6–9 Oct 2013, Santa Clara, CA, USA. IEEE (2013). http://ieeexplore.ieee.org/xpl/mostRecentIssue.jsp?punumber=6679357
31. Huang, Y., Zhang, C.: New data types and operations to support geo-streams. In: Cova, T.J., Miller, H.J., Beard, K., Frank, A.U., Goodchild, M.F. (eds.) GIScience. Lecture Notes in Computer Science, vol. 5266, pp. 106–118. Springer (2008)

32. Hunter, T., Das, T., Zaharia, M., Abbeel, P., Bayen, A.M.: Large-scale estimation in cyberphys-
    ical systems using streaming data: a case study with arterial traffic estimation. IEEE T. Autom
    Sci Eng **10**(4), 884–898 (2013)
33. ISO 19108:2002: Geographic information – Temporal schema (2002)
34. ISO 19107:2003: Geographic information – Spatial schema (2003)
35. ISO 19141:2008: Geographic information – Schema for moving features (2008)
36. ISO/IEC 13249-3:2011: Information technology – Database languages – SQL multimedia and
    application packages – Part 3: Spatial (2011)
37. Jiang, J., Bao, H., Chang, E.Y., Li, Y.: MOIST: a scalable and parallel moving object indexer
    with school tracking. PVLDB **5**(12), 1838–1849 (2012)
38. Kazemitabar, S.J., Kashani, F.B., McLeod, D.: Geostreaming in cloud. In: Ali, M.H., Hoel,
    E.G., Kashani, F.B. (eds.) Proceedings of the 2011 ACM SIGSPATIAL International Workshop
    on GeoStreaming, IWGS 2011, Nov 1, 2011, Chicago, IL, USA. pp. 3–9. ACM (2011). http://
    doi.acm.org/10.1145/2064959.2064962
39. Kornacker, M., Behm, A., Bittorf, V., Bobrovytsky, T., Ching, C., Choi, A., Erickson, J., Grund,
    M., Hecht, D., Jacobs, M., Joshi, I., Kuff, L., Kumar, D., Leblang, A., Li, N., Pandis, I.,
    Robinson, H., Rorke, D., Rus, S., Russell, J., Tsirogiannis, D., Wanderman-Milne, S., Yoder,
    M.: Impala: a modern, open-source SQL engine for Hadoop. In: CIDR 2015, Seventh Biennial
    Conference on Innovative Data Systems Research, Asilomar, CA, USA, Jan 4–7, 2015, Online
    Proceedings (2015). www.cidrdb.org
40. Lu, J., Güting, R.H.: Parallel SECONDO: practical and efficient mobility data processing in
    the cloud. In: Hu et al. [30], pp. 17–25. http://ieeexplore.ieee.org/xpl/mostRecentIssue.jsp?
    punumber=6679357
41. Ma, Q., Yang, B., Qian, W., Zhou, A.: Query processing of massive trajectory data based on
    MapReduce. In: Meng, X., Wang, H., Chen, Y. (eds.) Proceedings of the First International
    CIKM Workshop on Cloud Data Management, CloudDb 2009, Hong Kong, China, Nov 2,
    2009. pp. 9–16. ACM (2009). http://doi.acm.org/10.1145/1651263.1651266
42. Mahmood, A.R., Aly, A.M., Qadah, T., Rezig, E.K., Daghistani, A., Madkour, A., Abdelhamid,
    A.S., Hassan, M.S., Aref, W.G., Basalamah, S.: Tornado: a distributed spatio-textual stream
    processing system. PVLDB **8**(12), 2020–2031 (2015). http://www.vldb.org/pvldb/vol8/p2020-
    mahmood.pdf
43. Martin, D.: JTS Topology Suite (2016). http://tsusiatsoftware.net/jts/main.html
44. Meehan, J., Tatbul, N., Zdonik, S., Aslantas, C., Çetintemel, U., Du, J., Kraska, T., Madden,
    S., Maier, D., Pavlo, A., Stonebraker, M., Tufte, K., Wang, H.: S-store: streaming meets trans-
    action processing. PVLDB **8**(13), 2134–2145 (2015). http://www.vldb.org/pvldb/vol8/p2134-
    meehan.pdf
45. Miller, J., Raymond, M., Archer, J., Adem, S., Hansel, L., Konda, S., Luti, M., Zhao, Y.,
    Teredesai, A., Ali, M.H.: An extensibility approach for spatio-temporal stream processing
    using Microsoft StreamInsight. In: Pfoser, D., Tao, Y., Mouratidis, K., Nascimento, M.A.,
    Mokbel, M.F., Shekhar, S., Huang, Y. (eds.) SSTD. Lecture Notes in Computer Science, vol.
    6849, pp. 496–501. Springer (2011)
46. Mokbel, M.F., Xiong, X., Hammad, M.A., Aref, W.G.: Continuous query processing of spatio-
    temporal data streams in PLACE. GeoInformatica **9**(4), 343–365 (2005)
47. Murray, D.G., McSherry, F., Isaacs, R., Isard, M., Barham, P., Abadi, M.: Naiad: a timely
    dataflow system. In: Kaminsky, M., Dahlin, M. (eds.) ACM SIGOPS 24th Symposium on
    Operating Systems Principles, SOSP '13, Farmington, PA, USA, Nov 3–6, 2013. pp. 439–455.
    ACM (2013). http://dl.acm.org/citation.cfm?id=2517349
48. Neumeyer, L., Robbins, B., Nair, A., Kesari, A.: S4: distributed stream computing platform.
    In: Fan, W., Hsu, W., Webb, G.I., Liu, B., Zhang, C., Gunopulos, D., Wu, X. (eds.) ICDMW
    2010, The 10th IEEE International Conference on Data Mining Workshops, Sydney, Australia,
    13 Dec 2010. pp. 170–177. IEEE Computer Society (2010)
49. Nidzwetzki, J.K., Güting, R.H.: Distributed SECONDO: A highly available and scalable system
    for spatial data processing. In: Claramunt, C., Schneider, M., Wong, R.C., Xiong, L., Loh, W.,
    Shahabi, C., Li, K. (eds.) Advances in Spatial and Temporal Databases - 14th International

Symposium, SSTD 2015, Hong Kong, China, Aug 26–28, 2015. Proceedings. Lecture Notes in Computer Science, vol. 9239, pp. 491–496. Springer (2015). http://dx.doi.org/10.1007/978-3-319-22363-6

50. Oracle: Oracle Fusion Middleware – Oracle CQL Language Reference for Oracle Event Processing, 12c Release (12.1.3.0). Oracle Corporation (2014)

51. Patroumpas, K., Sellis, T.K.: Managing trajectories of moving objects as data streams. In: Sander, J., Nascimento, M.A. (eds.) STDBM. pp. 41–48 (2004)

52. Patroumpas, K., Sellis, T.K.: Event processing and real-time monitoring over streaming traffic data. In: Martino, S.D., Peron, A., Tezuka, T. (eds.) W2GIS. Lecture Notes in Computer Science, vol. 7236, pp. 116–133. Springer (2012)

53. Qian, Z., He, Y., Su, C., Wu, Z., Zhu, H., Zhang, T., Zhou, L., Yu, Y., Zhang, Z.: TimeStream: reliable stream computation in the cloud. In: Hanzálek, Z., Härtig, H., Castro, M., Kaashoek, M.F. (eds.) Eighth Eurosys Conference 2013, EuroSys '13, Prague, Czech Republic, April 14-17, 2013. pp. 1–14. ACM (2013)

54. Sarwat, M.: Interactive and scalable exploration of big spatial data - A data management perspective. In: Jensen, C.S., Xie, X., Zadorozhny, V., Madria, S., Pitoura, E., Zheng, B., Chow, C. (eds.) 16th IEEE International Conference on Mobile Data Management, MDM 2015, Pittsburgh, PA, USA, Vol. 1. pp. 263–270, June 15–18 2015. IEEE (2015). http://dx.doi.org/10.1109/MDM.2015.67

55. Schneider, M.: Spatial and spatio-temporal data models and languages. In: Liu, L., Özsu, M.T. (eds.) Encyclopedia of Database Systems, pp. 2681–2685. Springer, New York (2009)

56. Stonebraker, M., Çetintemel, U., Zdonik, S.B.: The 8 requirements of real-time stream processing. SIGMOD Rec. 34(4), 42–47 (2005)

57. Tan, H., Luo, W., Ni, L.M.: CloST: a Hadoop-based storage system for big spatio-temporal data analytics. In: Chen, X., Lebanon, G., Wang, H., Zaki, M.J. (eds.) 21st ACM International Conference on Information and Knowledge Management, CIKM'12, Maui, HI, USA, pp. 2139–2143 Oct 29–Nov 02, 2012. ACM (2012). http://doi.acm.org/10.1145/2396761.2398589

58. White, T.: Hadoop: The Definitive Guide, 2nd edn. O'Reilly Media, Inc., California (2012)

59. Yu, J., Wu, J., Sarwat, M.: GeoSpark: a cluster computing framework for processing large-scale spatial data. In: Proceedings of the 23rd SIGSPATIAL International Conference on Advances in Geographic Information Systems, pp. 70:1–70:4. GIS '15, ACM, New York, NY, USA (2015). http://doi.acm.org/10.1145/2820783.2820860

60. Zhang, C., Huang, Y., Griffin, T.: Querying geospatial data streams in SECONDO. In: Agrawal, D., Aref, W.G., Lu, C.T., Mokbel, M.F., Scheuermann, P., Shahabi, C., Wolfson, O. (eds.) GIS. pp. 544–545. ACM (2009)

61. Zheng, Y., Chen, Y., Li, Q., Xie, X., Ma, W.: Understanding transportation modes based on GPS data for web applications. TWEB 4(1) (2010). http://doi.acm.org/10.1145/1658373.1658374

62. Zheng, Y., Xie, X., Ma, W.: GeoLife: a collaborative social networking service among user, location and trajectory. IEEE Data Eng. Bull. 33(2), 32–39 (2010). http://sites.computer.org/debull/A10june/geolife.pdf

Chapter 4
Spatio-Temporal Data Stream Clustering

Abstract Spatio-temporal data streams are huge amounts of data pertaining to time and position of moving objects. Mining such amount of data is a challenging problem, since the possibility to extract useful information from this peculiar kind of data is crucial in many RFIP application scenarios. Moreover, spatio-temporal data streams pose interesting challenges for their proper representation, thus making the mining process harder than for classical data. In this chapter we deal with a specific spatio-temporal data stream class, namely *trajectory streams* that collect data pertaining to spatial and temporal position of mobile objects.

Keywords Knowledge discovery · Spatio-temporal data streams · Data stream clustering · Trajectory streams · Trajectory clustering

4.1 Introduction

We define knowledge discovery in data streams (KDDS) as a process of identifying valid, novel, useful, and understandable patterns from data streams. Data mining is the core of the KDDS process, involving the inferring of algorithms that explore the data streams, develop the model and discover previously unknown patterns—it is a process of extracting hidden knowledge structures represented in models and patterns of data streams. Discovery-oriented data stream mining methods automatically identify patterns in the data streams.

The goal of pattern recognition is the classification of objects into a number of categories or classes [41]. In what follows, we consider a specific class of pattern mining, i.e. pattern mining in the context of spatio-temporal streams produced by mobile objects. In other words, we will focus on relevant methodologies and algorithms, fundamental for on-line discovering groups of *trajectories* based on their proximity in spatio-temporal domain.

Taxonomy of movement patterns presented in [14] intends to contribute to the development of a toolbox of knowledge discovery algorithms by developing a conceptual framework for movement behaviour of different mobile objects and a comprehensive classification and review of movement patterns. This is indispensable as a basis for the development of pattern recognition which is required to be efficient,

© The Author(s) 2016
Z. Galić, *Spatio-Temporal Data Streams*, SpringerBriefs
in Computer Science, DOI 10.1007/978-1-4939-6575-5_4

effective and as generic as possible. A conceptual framework of the elements defining the spatio-temporal behavior of mobile objects and a comprehensive classification and definitions of movement patterns are developed and proposed.

The term *trajectory pattern* is generic term for a family of different pattern types that can be mined from trajectories of mobile objects.

Flexible pattern queries enable users to select trajectories based on specific events of interest. The queries are composed of a sequence of simple spatio-temporal predicates, as well as movement pattern predicates. *Density-based patterns* capture the aggregate behaviour of trajectories as groups [42].

According to the taxonomy of techniques for analyzing movement [4], clustering is a synoptic analyzing method for the tasks of gaining an overall concept and a concise description of a phenomenon, i.e. movement.

Most definitions of cluster are based on loosely defined terms, such as similar, and alike, etc., or they are oriented to a specific kind of cluster.

Definition 4.1 (Clustering) Let $X = \{x_1, x_2, \ldots, x_n\}$ be a data set. An *m-clustering* of X, $\mathbb{R}$, the partition of X into m sets (*clusters*), $C_1, \ldots, C_m$, so that the following three conditions hold:

- $C_i \neq \emptyset, i = 1, \ldots, m$
- $\bigcup_{i=1}^{m} C_i = X$
- $C_i \cap C_j = \emptyset, i \neq j, j = 1, \ldots, m$

In addition, the vectors contained in a cluster C_i are more similar to each other and less similar to the feature vectors of the other clusters. Quantifying the terms *similar* and *dissimilar* depends on the types of clusters involved.

4.1.1 Spatio-Temporal Clustering

Data clustering is one of the most important mining techniques exploited in the knowledge discovery process. Clustering spatio-temporal data in an unsupervised way is a challenging task—find a suitable partition achieving two dual goals: maximize the similarity of objects belonging to the same cluster and minimize the similarity among objects in different clusters. A number of different clustering methods have been developed in order to solve the problem from different perspective, i.e. *partition* based (*k-means* [34], *k-means++* [6], *k-means* ‖ [8]), *density* based (*DBSCAN* [17]), *hierarchical* (*BIRCH* [49]) and *grid-based* (*STING* [44]).

There is a number of interesting disk-based movement patterns: *flock* [9], *meet* [9], *leaders* and *followers* [3]. The disk-based movement patterns are satisfied by groups of objects that move together within a disc with user-specific predefined extend [25]. However, predefined disc shape and size may impose serious restrictions for many real applications. Density-based trajectory patterns have been introduced to overcome that restriction, yielding a group of clustered movement patterns: *moving cluster* [28], *convoy* [26], *swarm* [31] and *following* [45].

Density-based trajectory patterns follows widely accepted idea of density connection from *DBSCAN* algorithm [17], being able of to capture clusters of any extend, as long as the mobile objects meet predefined distance related constraint. Rather than produce clusters explicitly, *OPTICS* [5] generates an augmented cluster ordering representing density-based clustering structure [22], and has been used for trajectory clustering.

T-OPTICS [22, 35] focuses on two complementary goals: (i) to find the most adequate clustering method for trajectories, and (ii) to improve the quality of trajectory clustering by exploring the intrinsic semantic of the temporal dimension. The first goal has been achieved by adopting the *OPTICS* density-based clustering algorithm, and extending it by defining trajectories similarity criteria as the average distance between two trajectories. *T-OPTICS* delivers well defined trajectory clusters, with a good tolerance to noise and outliers. The second goal has been achieved by introducing the concept of *temporal focusing*, i.e. focusing on simple time intervals where the clustering structure of trajectories is semantically clearer than just considering the whole trajectories. An algorithm called *TF-OPTICS* generalizes *T-OPTICS* with a focus on the temporal dimension—basically enlarging the search space by considering the restrictions of the trajectories onto sub-intervals of time. *TF-OPTICS* detects the most meaningful time intervals, which allows to isolate the density-based clusters of higher quality.

TRACLUS algorithm [22, 25, 30] has been developed within partition-and-group framework, with the goal of discovering common sub-trajectories from trajectory database. Partitioning disaggregates each trajectory into a set of segments by adopting minimum description length (MDL) principle widely used in information theory. Grouping of trajectory segments that are closed to each other is done by line segment density based clustering algorithm, based on *DBSCAN* algorithm. It should be pointed out that *TRACLUS* generates the clusters of any shape and size, but does not consider the temporal aspects. As a consequence, some objects can belong to the same cluster even if they have completely different temporal attributes, and could be an appropriate method in applications which do not require the same temporal context, i.e. for discovering spatio-temporal patterns having lagged coincidence in time: hurricane forecasting, animal movements, etc.

M-Atlas [20] is the system and mobility data mining query language aimed at discovering trajectory patterns. M-Atlas explicitly supports four pattern types: *T-Cluster*, *T-Pattern*, *T-Flock* and *T-Flow*, as well as their aggregation and/or collection: Reachability plot, T-PTree and T-OD Matrix, which are the global models extracted by a data mining algorithm. Mastering the complexity of KDD process, based on the interleaving of an unsupervised method with a supervised one, could be expressed by combining query and mining functions. A clustering-based analytical process is interactive, involving the three steps: (i) clustering is performed over the sampled data set; (ii) one or more representative clusters are computed; and (iii) representatives are used to classify the entire data set.

In spite of extensive work on historical trajectories, only few works exist for clustering having roots in artificial intelligence and intelligent systems. *CenTra* [37] tackles the problem of discovering the centroid trajectory of a group of movements.

The method introduces intuitionistic fuzzy vector representation of trajectories, two-component distant metrics and a variant of *fuzzy k-means* [23] algorithm for clustering trajectories under uncertainty. Work presented in [11] is similar to that of [37]: both explore *fuzzy k-means* algorithm for trajectory clustering. A parallel spatial clustering algorithm *SPARROW* [18] is a multi-agent adaptive algorithm based on swarm intelligence techniques [21]. It combines the stochastic search of an adaptive flocking with the *DBSCAN* heuristics.

4.1.1.1 Distance Functions for Global and Local Similarity

Spatio-temporal clustering implies partition of spatio-temporal streams into clusters, such that each cluster contains similar trajectories according to some distance criteria. *Similarity* between trajectories may be considered in a number of different ways and depends on the application and goal of spatio-temporal pattern mining. Similarity is captured either by global or local *distance function*.

A global distance function defines the overall distance between two trajectories [25, 38]. A simple approach to measure between two trajectories $\mathcal{T}_1$ and $\mathcal{T}_2$ is to compute the sum of the Euclidean distance between all corresponding points in $\mathcal{T}_1$ and $\mathcal{T}_2$. However, this definition implies that $\mathcal{T}_1$ and $\mathcal{T}_2$ have the same length, which is not always true in the reality. The *geographic* global distance function [33] defines geographic similarity between two trajectories.[1]

The geographic distance between $\mathcal{T}_1$ and $\mathcal{T}_2$ is defined in the following way:

$$\mathcal{D}_{geo} = \overline{C_{\mathcal{T}_1}, C_{\mathcal{T}_2}} + \overline{C_{\mathcal{T}_1}, C_{\mathcal{T}_2}} \cdot \frac{|\,\|\mathcal{T}_1\| - \|\mathcal{T}_2\|\,|}{\max(\|\mathcal{T}_1\|, \|\mathcal{T}_2\|)} - \langle\|\Delta_{\mathcal{T}_1}\|, \|\Delta_{\mathcal{T}_2}\|\rangle \cdot \cos(\Delta_{\mathcal{T}_1}, \Delta_{\mathcal{T}_2})$$

(4.1)

where $\|\mathcal{T}_1\|$ and $\|\mathcal{T}_2\|$ denote the length, $C_{\mathcal{T}_1}$ and $C_{\mathcal{T}_2}$ denote center of mass, $\Delta_{\mathcal{T}_1}$ and $\Delta_{\mathcal{T}_2}$ denote displacement[2] of $\mathcal{T}_1$ and $\mathcal{T}_2$ respectively (Fig. 4.1).

The first term $\overline{C_{\mathcal{T}_1}, C_{\mathcal{T}_2}}$ measures the distance between the centers of mass. The second term measures the difference between the length of the trajectories. The third term (the average length of two trajectories times cosine similarity) reduces the distance between trajectories, where

$$\cos(\Delta_{\mathcal{T}_1}, \Delta_{\mathcal{T}_2}) = \frac{\Delta_{\mathcal{T}_1} \cdot \Delta_{\mathcal{T}_2}}{\|\Delta_{\mathcal{T}_1}\|\|\Delta_{\mathcal{T}_2}\|}$$

(4.2)

The cosine similarity between two vectors is a measure of the cosine of the angle between them, and here we use it to measure the cosine similarity between the displacements of $\|\mathcal{T}_1\|$ and $\|\mathcal{T}_2\|$.

---

[1] It could be also applied as a geographic similarity measure between two sub-trajectories.

[2] The shortest distance between the first an the last element/point.

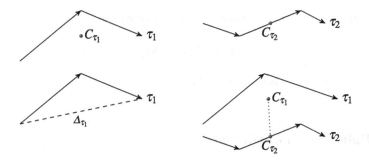

Fig. 4.1 Geographic distance components

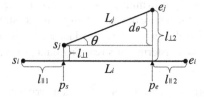

Fig. 4.2 Trajectory-Hausdorff distance components (*source* [30])

Global distance measure defines the overall similarity (i.e. distance) between two trajectories, but not their local similarity during some time interval. We define local distance function as Hausdorff distance:

$$\mathscr{D}_{Hausdorff} = w_\perp \cdot d_\perp + w_\| \cdot d_\| + w_\Theta \cdot d_\Theta \qquad (4.3)$$

where $w_\perp, w_\|$ and w_Θ are the application-specific weights of the distance components.

Figure 4.2 illustrates distance components[3] $d_\perp$ (perpendicular), $d_\|$ (parallel) and d_Θ (angle) between two line segments L_i and L_j. The line segments are defined by their starting and ending points s and e. We denote the projection points of s_j and e_j into L_i as p_s and p_e respectively.

The *perpendicular* distance $d_\perp$ between L_i and L_j is defined as

$$d_\perp = \frac{l_{\perp 1}^2 + l_{\perp 2}^2}{l_{\perp 1} + l_{\perp 2}} \qquad (4.4)$$

where $l_{\perp 1}$ and $l_{\perp 2}$ are two perpendicular distances between L_i and L_j.
The *parallel* distance $d_\perp$ between L_i and L_j is defined as

$$d_\| = \min\{l_{\|1}, l_{\|2}\} \qquad (4.5)$$

where $l_{\|1}$ and $l_{\|2}$ are two parallel distances between L_i and L_j.

---

[3] These components adapted from similarity measures used in the area of pattern recognition [41].

The *angular* distance d_Θ between L_i and L_j is defined as

$$d_\Theta = \|L_2\| \times sin\Theta \qquad\qquad (4.6)$$

where Θ is the angle between L_i and L_j.

4.2   Data Stream Clustering

Most data mining and knowledge discovery techniques assume a finite amount of persistent data and their analysis in multiple steps by a *batch* algorithm. However, due to constraints imposed by the nature of spatio-temporal data streams, this approach is not feasible. As opposed to clustering of a finite data set that is available entirely prior to the data mining process, clustering of streaming data poses additional challenges [29]:

1. *Single pass clustering*—Due to the fact that data arrive continuously, clustering has to be performed in a single pass over the data in an online fashion.
2. *Limited memory*—Since data streams are potentially unbounded, storing each arriving object is simply not feasible.
3. *Limited time*—The clustering algorithm has to be able to keep up with the speed of the data stream, i.e. clustering cannot take longer than the average time between any two objects in the stream.
4. *Varying time allowances*—The time available to process any item in the stream may vary greatly.
5. *Evolving data*—The model underlying the data in the stream may change over time. To capture such phenomena, stream clustering should be capable of detecting such changes. Denoted as *concept drift*, changes in clusters should be reported separately.
6. *Flexible number and size of clusters*—Many partitioning-based clustering algorithms require parametrization of the number of clusters to be detected. While setting such a parameter is also difficult in traditional clustering, setting a fixed number of clusters for the stream would distort the model.

Obviously, clustering data streams cannot be done using traditional algorithms for batch processing of persistent data sets.

The data stream clustering could be defined as *to maintain a continuously consistent good clustering of the sequence observed so far, using small amount of memory and time* [19].

Many data stream clustering algorithms rely on object-based paradigm which can be summarized into two main components: *online* component (also known as data abstraction step) and *offline* component (also known as clustering step), as illustrated in Fig. 4.3.

The online abstraction step summarizes the data stream with the help of particular data structures for dealing with space and memory constraints of stream applications.

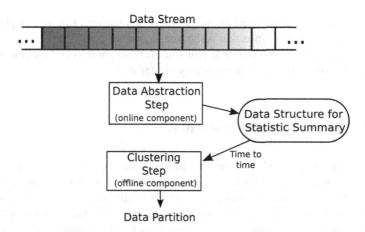

Fig. 4.3 Object-based data stream clustering framework (*source* [15])

These data structures summarize the stream in order to preserve the meaning of the original objects without the need of actually storing them. Designing suitable data structures (feature vector, prototype array, coreset tress, grids, etc.) is a crucial step, especially due to space-constraints assumptions in IFP applications based on centralized data stream processing architecture—the entire stream cannot be stored in the main memory.

As we already pointed out in Sect. 4.1.1, density-based clustering paradigm is of particular importance in spatio-temporal clustering. *DenStream* [10] belongs to the class of density-based data stream clustering algorithms. It uses two feature vectors based on *BIRCH* [49] clustering feature (CF) vector: *p-microsclusters* (potential cluster) and *o-microsclusters* (a buffer for supporting outlier detection).

Data stream clustering algorithms obtain a data partition via an offline clustering step (offline component). The offline component provides a quick understanding of the broad clusters in the data stream. Since this component requires the summary statistics as input, traditional batch clustering algorithms are used to find a data partition over the summaries. *DenStream* uses a *DBSCAN* variant for performing data stream clustering—the offline component receives the feature vectors in the form of *p-microsclusters* and two predefined parameters: boundary[4] (ε) and the integer weight of a given *p-microcluster* (μ).

The *ClusTree* is a self-adaptive anytime[5] clustering algorithm [29] which also uses a weighted clustering feature (CF) vector. The CFs are kept into a balanced, extended index structures from the R-tree family. Unlike the minimum bounding rectangles that R-tree family maintain (in addition to the objects), *ClusTree* stores

---

[4]Inserting an object into its nearest *p-microcluster* (or alternatively, *o-microcluster*) is successful if its updated radius r is within ε. For more details, please refer to [10].

[5]Clustering algorithms that are capable of delivering a result at any given point in time, and of using more time if it is available to refine the result [29].

only CFs. In order to adapt itself to exceptionally fast streams, *ClusTree* aggregates similar objects for a faster insertion in the tree.

It is worth to notice that both *DenStream* and *ClusTree* are density-based clustering algorithms capable to discover the arbitrary cluster shapes.

4.3 Trajectory Stream Clustering

Finding representative paths or common trends shared by mobile objects requires grouping of similar trajectories into clusters. A general strategy is to represent a trajectory with feature vector, denoting similarity between trajectories by the distance between their feature vectors. However, encoding the spatio-temporal properties of trajectory points into its feature vector is not trivial. Additionally, as different trajectories contain different and complex properties (shape, number of points, their temporal order, etc.), generating a feature vector with a uniform length is very challenging.

Trajectory clustering is a quite complex due to explicit temporal ordering of trajectory streams. There is a number of trajectory clustering methods and algorithms for persistent, historical trajectory data: *TRACLUS*, *T-Clustering*, *T-OPTICS*, etc. However, these methods and algorithms are not trajectory streams since clusters are re-calculated from scratch every time.

On the other hand, none of the existing data stream clustering algorithms can deal with trajectory streams. The reason for this is that each tuple in the data stream is an entry, while each typle in the trajectory stream is only a part of an entry.

4.3.1 Incremental Trajectory Clustering Using Micro- and Macro-Clustering

Let $\langle \mathscr{I}_{\tau_1}, \mathscr{I}_{\tau_2}, \ldots \rangle$ denote a sequence of time-stamped trajectory data sets, where each $\mathscr{I}_{\tau_i}$ is a set of trajectories being presented at time τ_i. Each $\mathscr{I}_{\tau i} = \{\mathscr{T}_1, \mathscr{T}_2, \ldots, \mathscr{T}_{n_\mathscr{T}}\}$ where each $\mathscr{T}_j$ is a trajectory represented by a series of temporally ordered points $p_1 \to p_2 \to \cdots \to p_n$, and each point p_i consists of a spatial coordinate set and a timestamp such that $p = \langle x_i, y_i, \tau_i \rangle$. Simplified trajectory $\mathscr{T}_j^s$ is represented as a sequence of connected directed line segments: $\mathscr{T}_j^s = \mathscr{L}_1, \mathscr{L}_2, \ldots \mathscr{L}_n$, where $\mathscr{L}_i$ and $\mathscr{L}_{i+1}$ are connected directed line segments (i.e., trajectory partitions).

The goal is to produce a set of clusters $\mathscr{O} = \{\mathscr{C}_1, \mathscr{C}_2, \ldots, \mathscr{C}_{n_\mathscr{C}}\}$. A cluster $\mathscr{C}$ is a set of directed trajectory line segments $\mathscr{C}_i = \{\mathscr{L}_1, \mathscr{L}_2, \ldots, \mathscr{L}_{n_i}\}$, where $\mathscr{L}_k$ is a directed line segment of simplified trajectory $\mathscr{T}_j^s$ at time τ_i.

TCMM adapts *CluStream* [1] micro-/macroclustering framework for trajectory data, similarly to object-based data stream clustering framework. Trajectory micro-clusters (Fig. 4.4b) are updated continuously and contain clusters at very fine granu-

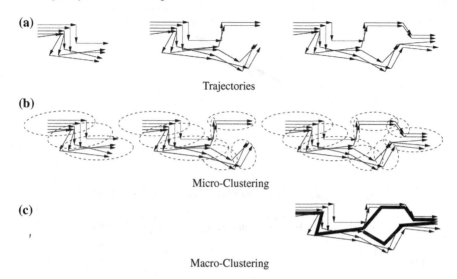

(a)

Trajectories

(b)

Micro-Clustering

(c)

Macro-Clustering

Fig. 4.4 The general framework and data flow of TCMM (*source* [32])

larity. They are restrictive in the sense that each micro-cluster holds and summarizes the information of local partitioned trajectories. The macro-clustering step receives the micro-clusters as input—it is only evoked after receiving the explicit request from the user.[6]

A trajectory micro-cluster $_\mu\mathscr{C}$ for a set of directed line segments $\{\mathscr{L}_1, \mathscr{L}_2, \ldots \mathscr{L}_n\}$ is tuple $(N, S_{center}, S_\theta, S_{length}, S^2_{center}, S^2_\theta, S^2_{length}, L_r)$, where

N —the number of line segments in the $_\mu\mathscr{C}$
S_{center} —the linear sum of of the line segments' center points
S_θ —the linear sum of of the line segments' angles
S_{length} —the linear sum of of the line segments' lengths
S^2_{center} —the squared sum of of the line segments' center points
S^2_θ —the squared sum of of the line segments' angles
S^2_{length} —the squared sum of of the line segments' lengths
L_r —the representative line segment.

The linear sum S represents the basic summarized information (center point, angle and length) of line segments, whereas the square sum S^2 is used to determine the *tightness* of micro-cluster.

The representative line segment (Fig. 4.5) L_r of $_\mu\mathscr{C}$ is defined by the starting point s_{L_r} and ending point e_{L_r}:

---

[6]Although this method handles only incremental data, it can be extended for trajectory streaming.

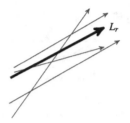

Fig. 4.5 Representative line segment

$$s_{L_r} = (x_{center} - \frac{\cos \theta}{2} \cdot length, \ y_{center} - \frac{\sin \theta}{2} \cdot length)$$
$$e_{L_r} = (x_{center} + \frac{\cos \theta}{2} \cdot length, \ y_{center} + \frac{\sin \theta}{2} \cdot length) \tag{4.7}$$

where

$$x_{center} = S_{center_x}/N$$
$$y_{center} = S_{center_y}/N$$
$$length = S_{length}/N \tag{4.8}$$
$$\theta = S_\theta/N$$

4.3.1.1 Distance Between Line Segment and Micro-Cluster

The distance between a line segment L_j and a micro-cluster $_\mu \mathscr{C}_k$ is in fact defined between line segment L_j and representative segment L_r of $_\mu \mathscr{C}_k$. The distance is an adapted line segment Hausdorff distance Fig. 4.2 and uses the center point distance $d_c{}^7$ instead of perpendicular distance $d_\perp$.

We define the longer line segment L_l and the shorter line segment L_s as:

$$L_l = L_i \quad | \quad \|L_i\| = max(\|L_j\|, \|L_r\|)$$
$$L_s = L_i \quad | \quad \|L_i\| = min(\|L_j\|, \|L_r\|) \tag{4.9}$$

The distance function is defined as:

$$\mathscr{D}_{TCMM}(L_l, L_s) = d_c + d_\| + d_\Theta \tag{4.10}$$

where d_c is the Euclidean distance between center points of L_l and L_s:

$$d_c = \|C_l - C_s\| \tag{4.11}$$

The parallel distance (Fig. 4.2):

---

[7] Similarly to *center of mass* of geographic distance, defined by 4.1.

Algorithm 4 TCMM—Trajectory Clustering using Micro- and Macro-clustering

---

Input: $\mathscr{I}_{now} = \{\mathscr{T}_1, \mathscr{T}_2, \ldots, \mathscr{T}_{n_{\mathscr{T}}}\}$,   ▷new trajectories
      $_\mu\mathscr{C} = \{_\mu\mathscr{C}_1, _\mu\mathscr{C}_2, \ldots, _\mu\mathscr{C}_{n_{\mu\mathscr{C}}}\}$ ▷existing micro-clusters
Parameters: δ_{max},   ▷distance threshold
      $\mathscr{M}$ ▷memory space constraint
  for all $\mathscr{T}_i \in \mathscr{I}_{now}$ **do**
    for all $\mathscr{L}_j \in \mathscr{T}_i$ **do**
      $_\mu\mathscr{C}_k \leftarrow \min(\mathscr{D}_{TCMM}(\mathscr{L}_j, _\mu\mathscr{C}))$
      if $\mathscr{D}_{TCMM}(\mathscr{L}_j, _\mu\mathscr{C}_k) \leq \delta_{max}$ **then**
        $_\mu\mathscr{C}_k \leftarrow_\mu\mathscr{C}_k \oplus L_j$
        $update(N, S_{center}, S_\theta, S_{length}, S^2_{center}, S^2_\theta, S^2_{length}, L_r)$
      else
        new $_\mu\mathscr{C}_{new}$ ▷create a new empty micro-cluster
        $_\mu\mathscr{C}_{new} \leftarrow_\mu\mathscr{C}_{new} \oplus L_j$
        $update(N, S_{center}, S_\theta, S_{length}, S^2_{center}, S^2_\theta, S^2_{length}, Lr)$
        $_\mu\mathscr{C} \leftarrow_\mu\mathscr{C} \oplus_\mu\mathscr{C}_{new}$
        if $\mathscr{M}_{\mu\mathscr{C}} > \mathscr{M}$ **then**   ▷merge micro-clusters in $_\mu\mathscr{C}$
          for $i \leftarrow 1, n$ **do**
            for $j \leftarrow i + 1, n$ **do**
              $\mathscr{D}^*_{TCMM_{ij}} \leftarrow \mathscr{D}^*_{TCMM}(_\mu\mathscr{C}_i, _\mu\mathscr{C}_j)$
            end for
          end for
          sort $\mathscr{D}^*_{TCMM}$
          repeat
            $_\mu\mathscr{C}_i \leftarrow_\mu\mathscr{C}_i \oplus_\mu\mathscr{C}_j$ ▷merge the most similar pairs
            $_\mu\mathscr{C}_j \leftarrow \varnothing$
          until $\mathscr{M}_{\mu\mathscr{C}} > \mathscr{M}$
        end if
      end if
    end for
  end for
  return $_\mu\mathscr{C}$

$$d_\| = \min(l_{\|1}, l_{\|2}) \qquad (4.12)$$

The angle distance (Fig. 4.2):

$$d_\theta = \begin{cases} \|L_s\| \times \sin\theta, & 0 \leq \theta < \pi/2 \\ \|L_s\|, & \pi/2 \leq \theta \leq \pi \end{cases} \qquad (4.13)$$

4.3.1.2   Merging Micro-Clusters

If the total memory space used by $_\mu\mathscr{C}$ exceeds a given memory space constraint $\mathscr{M}$, some micro-clusters have to be merged. At the same time, if the number of micro-clusters keeps increasing, it will influence the algorithm efficiency because finding the nearest micro-cluster is the most time-consuming task. Additionally, it is

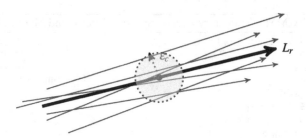

Fig. 4.6 Micro-cluster *center* extent

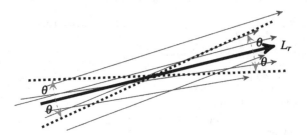

Fig. 4.7 Micro-cluster θ extent

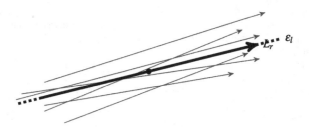

Fig. 4.8 Micro-cluster *length* extent

not necessary to keep all the micro-clusters since some of them may become closer after several updates.

We will use micro-cluster *extend* to define the distance between micro-clusters. The extend of the micro-cluster ε is triple $(\varepsilon_c, \varepsilon_\theta, \varepsilon_l)$ to measure the tightness of three basic features of a trajectory micro-cluster:

$$\varepsilon_\alpha = \sqrt{\frac{N \times S_\alpha^2 - (S_\alpha)^2}{N^2}} \tag{4.14}$$

where α represents c, θ or l (Figs. 4.6, 4.7 and 4.8, respectively.)

The purpose of extend is to fine-tune the distance function based on the tightness of micro-clusters. Extend is used to decrease the distance between the representative line segments of micro-clusters and enable loose micro-clusters to be more easily merged and vice-versa.

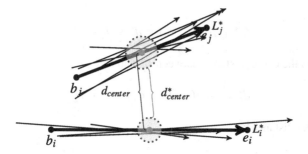

Fig. 4.9 *Center* distance with extend

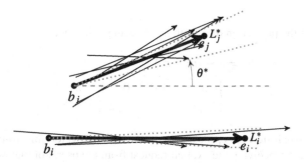

Fig. 4.10 *Angle* distance with extend

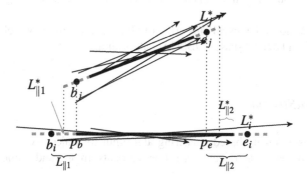

Fig. 4.11 *Parallel* distance with extend

The distance between micro-cluster $_\mu\mathscr{C}_i$ and $_\mu\mathscr{C}_j$ is equivalent to the distance between the representative line segments $L^*_{_\mu\mathscr{C}_i}$ with extend $\varepsilon_{_\mu\mathscr{C}_i}$ and $L^*_{_\mu\mathscr{C}_i}$ with extend $\varepsilon_{_\mu\mathscr{C}_j}$.

Figures 4.9, 4.10 and 4.11 shows an intuitive example of micro-cluster distance measure with extend. The distance between the centers (Fig. 4.9) is the distance between representative line segments minus the center extends of two micro-clusters.

The distance between two micro-clusters is modified distance function (4.10) between line segment and micro-cluster:

$$\mathscr{D}^*_{TCMM}(\mu\mathscr{C}_i, \mu\mathscr{C}_j) = d^*_c + d^*_\parallel + d^*_\Theta \tag{4.15}$$

where d_c is the center distance between L^*_i and L^*_j:

$$d_c = max(0, \|C_i - C_j\| - \varepsilon^i_c - \varepsilon^j_c) \tag{4.16}$$

The parallel distance[8]:

$$d^*_\parallel = max(0, min(l_{\parallel 1}, l_{\parallel 2}) - \frac{\varepsilon^i_l + \overline{\varepsilon^j_l}}{2}), \tag{4.17}$$

where $\overline{\varepsilon^j_l}$ is the projection of ε^j_l onto L^*_i. The angle distance:

$$\theta^* = \theta - (\varepsilon^i_\theta + \varepsilon^j_\theta)$$
$$d^*_\theta = \begin{cases} \|L^*_j\| \times \sin\theta^*, & 0 \leq \theta < \pi/2 \\ \|L_j\|, & \pi/2 \leq \theta \leq \pi \end{cases} \tag{4.18}$$

While micro-clustering is processed continuously, trajectory macro-clustering is evoked only after receiving the explicit request from the user. For that purpose, it is possible to adapt any density-based clustering algorithm by replacing the distance between spatial points with the distance between micro-clusters $\mathscr{D}^*_{TCMM}$ as defined in 4.15.

Figure 4.12 shows the visual clustering of GPS data recorded by mobile objects, using TCMM and density-based DBSCAN algorithm.

4.3.2   CTraStream

CTraStream is a density-based clustering algorithm for trajectory data streams. It considers the incremental trajectories as line segments streams and applies two-stage strategy:

(i) *CLnStream*—trajectory line segment stream clustering to obtain clusters of current time interval
(ii) *TraCluUpdate*—online update of trajectory clusters and extracting closed trajectory clusters based on the TC-tree structure

First, we define line segment distance between two mobile objects during the same time interval as (Fig. 4.13):

---

[8]In this case, $L^*_{\parallel 1} = 0$.

(a)

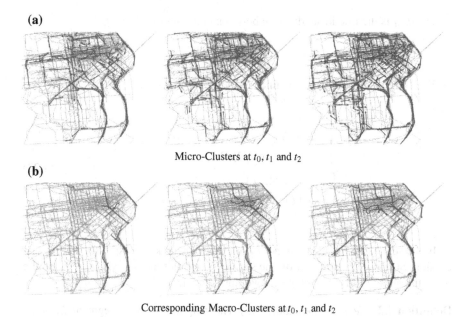

Micro-Clusters at t_0, t_1 and t_2

(b)

Corresponding Macro-Clusters at t_0, t_1 and t_2

Fig. 4.12 Micro- and Macro-Clusters in TCMM (*source* [32])

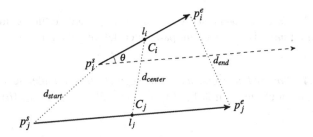

Fig. 4.13 Distance between *line* segments and the included angle θ

$$\mathscr{D}_{CTraStream}(l_i, l_j) = \alpha \times d_{start}(l_i, l_j) + \beta \times d_{center}(l_i, l_j) + \gamma \times d_{end}(l_i, l_j)$$
$$(4.19)$$

where d_{start} is the Euclidean distance between the starting point p_i^s of segment l_i and the starting point p_j^s of segment l_i and l_j:

$$d_{start} = \|p_i^s - p_j^s\| \qquad (4.20)$$

The d_{center} is the Euclidean distance between center points of l_i and l_j:

$$d_{center} = \|C_i - C_j\| \qquad (4.21)$$

The d_{end} is the Euclidean distance between end points of l_i and l_j:

$$d_{end} = \|p_i^e - p_j^e\| \tag{4.22}$$

The weights α, β and γ are defined in the following way:

$$
\begin{cases}
\begin{cases}
\alpha = (1 - \sin\theta/2)/3 \\
\beta = 1/3 \\
\gamma = (1 + \sin\theta/2)/3
\end{cases} , 0 \leq \theta \leq \pi/2 \\[2mm]
\begin{cases}
\alpha = (\sin\theta/2)/3 \\
\beta = 1/3 \\
\gamma = (2 - \sin\theta/2)/3
\end{cases} , \pi/2 < \theta \leq \pi
\end{cases}
\tag{4.23}
$$

In the sequel we define a number of relevant concepts necessary for further discussion. Suppose $\mathscr{L}$ is the set of line segments within time interval $[\tau_i, \tau_{i+1}]$ and ε is the threshold of distance between two line segments.

Definition 4.2 (*Segment neighborhood*) The neighborhood of segment l_i, is set denoted by $\mathscr{N}_{\mathscr{L}}(l_i) = \{l | l \in \mathscr{L} \wedge \mathscr{D}_{CTraStream}(l, l_j) \leq \varepsilon\}$. We denote the cardinality of $\mathscr{N}_{\mathscr{L}}(l_i)$ as $\|\mathscr{N}_{\mathscr{L}}(l_i)\|$.

Definition 4.3 (*Core line segment*) A line segment l_i is a core line segment l_{core} if $\mathscr{N}_{\mathscr{L}}(l_i) \geq min_{ls}$; min_{ls} is an important parameter defining the minimum number of line segments.

Definition 4.4 (*Border line segment*) A line segment l_i is a border segment l_{border} of the line segment cluster $\mathscr{C}$ if $\|\mathscr{N}_{\mathscr{L}}(l_i)\| < min_{ls}, \exists l_{core} | l_{core} \in \mathscr{N}_{\mathscr{L}}(l_i)$, such that $l_{core} \in \mathscr{C}$.

Definition 4.5 (*Isolated line segment*) A line segment l_i is an isolated segment only if it is neither a core line segment nor a border line segment.

Definition 4.6 (*Directly density-reachable segment*) A line segment l_i is directly density-reachable from l_{core} if $l_{core} \in \mathscr{N}_{l_i}$.

Definition 4.7 (*Density-reachable segment*) A line segment l_i is density-reachable from a line segment l_j, if there is a series of line segments $l_i, l_{i+1}, \ldots l_{j-1}, l_j \in \mathscr{L}$ such that l_k is directly-reachable from $l_{k+1} (k = i, i + 1, \ldots, j - 1, j)$.

Definition 4.8 (*Density-connected*) If two line segments l_i and l_j are both density-reachable from a line segment l_k, then l_i and l_j are mutually density-connected.

Definition 4.9 (*Line-segment-cluster*) A line-segment-cluster $\mathscr{C}$ is a non-empty set of line segments which are density-connected to each other and satisfy the *connectivity* and *maximality* conditions:

1. Connectivity: $\forall l_i; l_j \in \mathscr{C}$, l_i is a density-connected to l_j
2. Maximality: $\forall l_i; l_j \in \mathscr{L}$, if $l_i \in \mathscr{C}$ and l_j is density-reachable from l_i, then $l_j \in \mathscr{C}$.

Each line-segment-cluster $\mathscr{C}$ contains at least one core line segment l_{core} and all line segments being density-reachable from one of the core line segments. We denote the set of line-segment-clusters in time interval $[\tau_i, \tau_{i+1}]$ as
$$\mathscr{CL}_{[\tau_i, \tau_{i+1}]} = \{\mathscr{C}^1_{[\tau_i, \tau_{i+1}]}, \mathscr{C}^2_{[\tau_i, \tau_{i+1}]}, \ldots, \mathscr{C}^k_{[\tau_i, \tau_{i+1}]}\}.$$

Figure 4.14 shows trajectory streams of 18 mobile objects. Each line segment with an arrow is marked with the time interval and stands for a trajectory line segment. All trajectory line segments in a gray ellipse constitute a line-segment-cluster.[9]

Definition 4.10 (*Trajectory cluster*) We denote a trajectory cluster $\mathscr{C}_{\mathscr{T}}$ for mobile objects in the trajectory stream $\mathscr{T}_{\mathscr{S}}$ as a 2-tuple $(\mathcal{O}_m, [\tau_i, \tau_j])$, where $\mathcal{O}_m$ is the set of mobile objects in the trajectory cluster, τ_i and τ_j is the start and end time of the trajectory cluster respectively:
$$\mathscr{C}_{\mathscr{T}} : \exists \mathscr{C}^{i_k}_{[\tau_k, \tau_{k+1}]} \mid \mathscr{C}^{i_k}_{[\tau_k, \tau_{k+1}]} \in \mathscr{CL}_{[\tau_k, \tau_{k+1}]}, \mathcal{O}_m \subseteq \bigcap_{k=i}^{j-1} \mathscr{C}^{i_k}_{[\tau_k, \tau_{k+1}]} (\|\mathcal{O}_m\| \geq min_{\mathcal{O}}),$$
$$k=i,\ldots,j-1,$$
and $min_{\mathcal{O}}$ is object set number threshold.

Definition 4.11 (*Object-closed trajectory cluster*)
$$\mathscr{C}^{oc}_{\mathscr{T}} : \exists (\mathcal{O}'_m, [i,j]) \mid \mathcal{O}'_m \supset \mathcal{O}^{oc}_m \wedge \neg \mathscr{C}'_{[\tau_i, \tau_{i+1}]}$$

Definition 4.12 (*Time-closed trajectory cluster*)
$$\mathscr{C}^{tc}_{\mathscr{T}} : \exists (\mathcal{O}_m, [i-1, i]), (\mathcal{O}_m, [j, j+1]) \mid \neg \mathscr{C}_{[\tau_{i-1}, \tau_i]}, \neg \mathscr{C}_{[\tau_j, \tau_{j+1}]}$$

Definition 4.13 (*Closed trajectory cluster*) A trajectory cluster is called a closed trajectory cluster $\mathscr{C}^c_{\mathscr{T}}$, if and only if is both object-closed and time-closed:
$$\mathscr{C}^c_{\mathscr{T}} \leftrightarrow (\mathscr{C}^{oc}_{\mathscr{T}} \wedge \mathscr{C}^{tc}_{\mathscr{T}})$$

The line-segment-clusters $\mathscr{CL}_{[\tau, \tau+1]}$ of continuous time intervals constitute a closed trajectory cluster (Fig. 4.14).

A general framework of CTraStream is described in Algorithm 5:

Algorithm 5 CTraStream—Clustering Trajectory Stream

Input: $\mathscr{S}_{\mathscr{T}}$, ▷trajectory stream
  $TC - tree$, ▷closed trajectory cluster binary tree
  L ▷list of current closed trajectory clusters
Parameters: ε, ▷distance threshold between two segments
  min_{ls} ▷minimum number of line segments in line-segment-cluster
$\mathscr{CL} \leftarrow \varnothing$
for all $p_\tau \in \mathscr{S}_{\mathscr{T}}$ **do**
  $l_{new} \leftarrow \overrightarrow{p_{\tau-1}p_\tau}$
  $\mathscr{CL}_{[\tau-1, \tau]} \leftarrow CLnStream(l_{new}, \mathscr{CL}, \varepsilon, min_{ls})$ ▷Algorithm 6
  $TC - tree \leftarrow TraCluUpdate(\mathscr{CL}_{[\tau-1, \tau]}, TC - tree, L, min_{ls})$ ▷Algorithm 7
end for
return $TC - tree$

---

[9]In this case, the minimum number of line segments (min_{ls}) is 2.

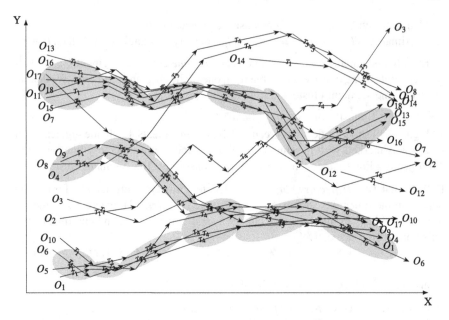

Fig. 4.14 Line segment clusters and trajectory clusters

4.3.2.1 CLnStream: Incremental Line-Segment Clustering

CLnStream performs clustering for line segments of current time interval $[\tau - 1, \tau]$ and gets line-segment-clusters $\mathscr{C}^i$ at the end of the current time interval. For each new point p_τ a new line segment $l_{new} = \overrightarrow{p_{\tau-1}p_\tau}$ is created and added to the line segments set $\mathscr{L}$ of the time interval $[\tau - 1, \tau]$. The new line segment l_{new} only influences line-segment-clusters of current time interval $\mathscr{C}_{\mathscr{L}_{[\tau-1,\tau]}}$, and we only need to perform the update processing on l_{new} and its neighborhood. The effected line segments could be classified into two categories:

New core line segments: some border line segments l_{border} may become a core line segment l_{core} and the corresponding clusters, in which these line segments are, will be updated. Let $\mathscr{L}_{core}^{new}(l_new)$ denote set of new core line segments:

$$\mathscr{L}_{core}^{new}(l_new) = \{l \mid \|\mathscr{N}_\mathscr{L}(l_i)\| \geq min_{ls} \wedge \|\mathscr{N}_{\mathscr{L}-l_new}\| < min_{ls}\} \qquad (4.24)$$

Updated core line segments: due to the clusters where the new line core line segments are to be updated, the previous core line segments within neighborhood of $\mathscr{L}_{core}^{new}(l_{new})$ need to be updated as well. Let $\mathscr{L}_{core}^{update}(l_{new})$ denote set of new core line segments:

$$\mathscr{L}_{core}^{update}(l_{new}) = \{l \mid \exists l_c \in \mathscr{L}_{core}^{new}(l_{new}), l \in \mathscr{N}_{\mathscr{L}-l_{new}}(l_c) \wedge \|\mathscr{N}_{\mathscr{L}-l_{new}}(l)\| \geq min_{ls}\}$$
$$(4.25)$$

If l_{new} is a core line segment, all segments in $\mathscr{L}_{core}^{new}(l_{new})$ and $\mathscr{L}_{core}^{update}(l_{new})$ are density-reachable from l_{new}, and all line-segment-clusters $\mathscr{N}_{\mathscr{L}}(l_{new})$ within neighborhood of l_{new} should be merged into one line-segment-cluster. Otherwise, there can exist multiple line-segment-clusters in the neighborhood of l_{new}. Therefore, if l_{new} is a core line segment, we merge all local line-segment-clusters where $\mathscr{L}_{core}^{update}(l_{new})$ are merged into one line-segment-cluster, and ingest $\mathscr{L}_{core}^{new}(l_{new})$ and all line segments being directly density-reachable from them into the line-segment-cluster. Otherwise, we have to process each element of $\mathscr{L}_{core}^{new}(l_{new})$ like the processing of l_{new} being a core line segment.

Algorithm 6 CLnStream—Clustering Line Segment Stream

Input: l_{new},   ▷new line segment

    $\mathscr{C}_{\mathscr{L}[\tau-1,\tau]} = \{\mathscr{C}_{[\tau-1,\tau]}^1, \mathscr{C}_{[\tau-1,\tau]}^2, \ldots, \mathscr{C}_{[\tau-1,\tau]}^k\}$,   ▷existing line clusters

    L,   ▷list of current closed trajectory clusters

    ε,   ▷distance threshold between two segments

    min_{ls} ▷minimum number of line segments in line-segment-cluster

find $\mathscr{L}_{core}^{new}(l_{new})$

find $\mathscr{L}_{core}^{update}(l_{new})$

if $l_{new} \in \mathscr{L}_{core}^{new}(l_{new})$ **then**

    if $\|\mathscr{C}_{\mathscr{L}_{core}^{update}(l_{new})}\| = 0$ **then**

        new $\mathscr{C} \leftarrow \{\mathscr{L}_{core}^{new}(l_{new}) \cup \mathscr{N}_{\mathscr{L}}(\mathscr{L}_{core}^{new}(l_{new}))\}$

        $\mathscr{C}_{\mathscr{L}[\tau-1,\tau]} \leftarrow \mathscr{C}_{\mathscr{L}[\tau-1,\tau]} \cup \{\mathscr{C}\}$

    else if $\|\mathscr{C}_{\mathscr{L}_{core}^{update}(l_{new})}\| = 1$ **then**

        $\mathscr{C}_{[\tau-1,\tau]}^t \leftarrow \mathscr{C}_{[\tau-1,\tau]}^t \cup \{\mathscr{L}_{core}^{new}(l_{new}) \cup \mathscr{N}_{\mathscr{L}}(\mathscr{L}_{core}^{new}(l_{new}))\}$

    else if $\|\mathscr{C}_{\mathscr{L}_{core}^{update}(l_{new})}\| > 1$ **then**

        $n \leftarrow \|\mathscr{C}_{\mathscr{L}_{core}^{update}(l_{new})}\|$

        $\mathscr{C}_{[\tau-1,\tau]}^t \leftarrow \mathscr{C}_{[\tau-1,\tau]}^t \cup \mathscr{C}_{[\tau-1,\tau]}^{t_1} \cdots \mathscr{C}_{[\tau-1,\tau]}^{t_n} \cup \{\mathscr{L}_{core}^{new}(l_{new}) \cup \mathscr{N}_{\mathscr{L}}(\mathscr{L}_{core}^{new}(l_{new}))\}$

    end if

else

    for all $l_q \in \mathscr{L}_{core}^{new}(l_{new})$ **do**

        find $\mathscr{L}_{core}^{new}(l_q)$

        if $\|\mathscr{N}_{\mathscr{L}core}(l_q)\| = 0$ **then**

            new $\mathscr{C} \leftarrow \{l_q \cup \mathscr{N}_{\mathscr{L}}(l_q)\}$

            $\mathscr{C}_{\mathscr{L}[\tau-1,\tau]} \leftarrow \mathscr{C}_{\mathscr{L}[\tau-1,\tau]} \cup \{\mathscr{C}\}$

        else if $\|\mathscr{N}_{\mathscr{L}core}(l_q)\| = 1$ **then**

            $\mathscr{C}_{[\tau-1,\tau]}^t \leftarrow \mathscr{C}_{[\tau-1,\tau]}^t \cup \{l_q \cup \mathscr{N}_{\mathscr{L}}(l_q)\}$

        else if $\|\mathscr{N}_{\mathscr{L}core}(l_q)\| > 1$ **then**

            $n \leftarrow \|\mathscr{N}_{\mathscr{L}core}(l_q)\|$

            $\mathscr{C}_{[\tau-1,\tau]}^t \leftarrow \mathscr{C}_{[\tau-1,\tau]}^t \cup \mathscr{C}_{[\tau-1,\tau]}^{t_1} \cdots \mathscr{C}_{[\tau-1,\tau]}^{t_n} \cup \{l_q \cup \mathscr{N}_{\mathscr{L}}(l_q)\}$

        end if

    end for

end if

return $\mathscr{C}_{\mathscr{L}[\tau-1,\tau]}$

4.3.2.2 TraCluUpdate

TraCluUpdate is responsible for online update trajectory clusters $\mathscr{C}_{\mathscr{S}\mathscr{T}}$ and extract closed trajectory clusters $\mathscr{C}^c_{\mathscr{S}\mathscr{T}}$. It uses *TC-tree*—a LCRS[10] binary tree with multiple root nodes. Each non-root node represents a closed trajectory cluster, whereas left sub-tree of each root node contains a series of closed trajectory clusters with the same start time. Figure 4.15 shows an example of *TC-tree*[11] during time interval $[\tau_0, \tau_2]$. *TraCluUpdate* also manages a single linked list L indexing all current closed trajectory clusters that will be updated in next time interval.

For the set of line-segment-clusters $\mathscr{C}_{\mathscr{L}_{[\tau_j, \tau_{j+1}]}}$ in current time interval $[\tau_j, \tau_{j+1}]$ and the node $node_{TC} = (\mathscr{O}_m, [\tau_j, \tau_{j+1}])$, if its left child-node is not null, then its left child-node $(\mathscr{O}'_m, [\tau_j, \tau_{j+1}])$ satisfies the condition that $\exists \mathscr{C}^{ik}_{[\tau_j, \tau_{j+1}]}$ makes $\mathscr{O}'_m = \mathscr{O}_m \cap \mathscr{C}^{ik}_{[\tau_j, \tau_{j+1}]}$.

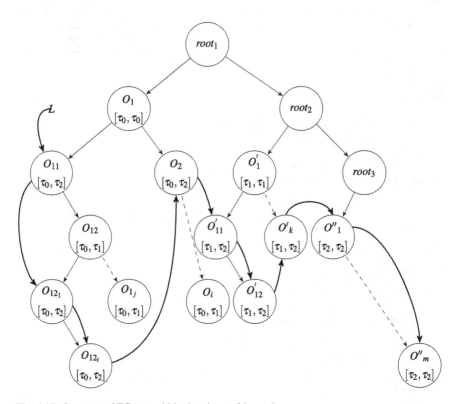

Fig. 4.15 Structure of TC-tree within time interval $[\tau_0, \tau_2]$

---

[10]Left Child-Right Sibling.

[11]O_{12_1} and O_{12_t} stands for a subset of O_{12}. Thus the time interval of node of O_{12_1} must be larger than that of node of O_{12}, denoted by $[\tau_0, \tau_2]$. O_{1_j} denotes j^{th} subset of O_1, and it also means that there are j subsets generated by O_1 that intersects with clusters of τ_2 time instance. O_{12_t} also denotes t^{th} subset of O_{12}; i, k and m has same meaning with j and t.

If the right child of its left child-node is not null, then the right child-node of its left child-node ($\mathscr{O}''_m$, $[\tau_j, \tau_{j+1}]$) satisfies the conditions of $\mathscr{O}''_m = \mathscr{O}_m \cap \mathscr{C}^{ik}_{[\tau_j, \tau_{j+1}]}$, and $\mathscr{O}'_m \cap \mathscr{O}''_m = \varnothing$. If the left sub-tree of $node_{TC}$ is null, then $\forall \mathscr{C}^{ik}_{[\tau_j, \tau_{j+1}]} \in \mathscr{C}^{ik}_{\mathscr{L}_{[\tau_j \cdot \tau_{j+1}]}}$, ($\mathscr{O}_m \cap \mathscr{C}^{ik}_{[\tau_j, \tau_{j+1}]}$, $[i, j+1]$) is not a closed trajectory cluster.

TC-tree poses the following two properties:

1. For a node ($\mathscr{O}_m$, $[\tau_{begin}, \tau_{end}]$) of *TC-tree*, any non-null node ($\mathscr{O}'_m$, $[\tau'_{begin}, \tau'_{end}]$) in its left sub-tree satisfies condition: $\mathscr{O}'_m \subset \mathscr{O}_m \wedge \tau'_{begin} = \tau_{begin} \wedge \tau'_{end} > \tau_{end}$.
2. For a node ($\mathscr{O}_m$, $[\tau_{begin}, \tau_{end}]$) of *TC-tree*, any non-null node ($\mathscr{O}'_m$, $[\tau'_{begin}, \tau'_{end}]$) in its right sub-tree satisfies condition: $\mathscr{O}'_m \cap \mathscr{O}_m = \varnothing \wedge \tau'_{begin} = \tau_{begin}$.

TC-tree is update according to the following rules:

R_1: Consider inserting a new node $node^{new}_{TC}$ into sub-tree of *TC-tree* (a node $node_{TC}$ of *TC-tree*). For left sub-tree inserting, if the left child of $node_{TC}$ is null, then insert $node^{new}_{TC}$ into its left child, otherwise insert $node^{new}_{TC}$ into the right sub-tree of its left child. For right sub-tree inserting, if the right child of $node_{TC}$ is null, then insert $node^{new}_{TC}$ into its right child, otherwise insert $node^{new}_{TC}$ into the right sub-tree of its right-child.

R_2: Consider a node $node_{TC} = (\mathscr{O}_m, [\tau_{begin}, \tau_j])$ of L: If $\exists \mathscr{C}^i_{[\tau_j, \tau_{j+1}]} \mid \mathscr{O}_m \cap \mathscr{C}^i_{[\tau_j, \tau_{j+1}]} \supseteq \mathscr{O}_m$, $1 \leq i \leq k$, then update $node_{TC}$ to ($\mathscr{O}_m$, $[\tau_{begin}, \tau_{j+1}]$).

R_3: $\forall(\mathscr{O}_m, [\tau_{begin}, \tau_j]) \in L$ if $\exists \mathscr{C}^i_{[\tau_j, \tau_{j+1}]}(1 \leq i \leq k) \mid \|\mathscr{O}_m \cap \mathscr{C}^i_{[\tau_j, \tau_{j+1}]}\| \geq min_{ls} \wedge \mathscr{O}_m \cap \mathscr{C}^i_{[\tau_j, \tau_{j+1}]} \neq \varnothing$ then backward check whether the trajectory cluster ($\mathscr{O}_m$, $[\tau_{begin}, \tau_{j+1}]$) is closed. If closed, insert the trajectory cluster ($\mathscr{O}_m$, $[\tau_{begin}, \tau_{j+1}]$) into the left sub-tree of ($\mathscr{O}_m$, $[\tau_{begin}, \tau_j]$), and insert the ($\mathscr{O}_m$, $[\tau_{begin}, \tau_{j+1}]$) into L at next of ($\mathscr{O}_m$, $[\tau_{begin}, \tau_j]$). Update ($\mathscr{O}_m$, $[\tau_{begin}, \tau_j]$) and remove it from L.

R_4: $\forall \mathscr{C}^i_{[\tau_j, \tau_{j+1}]}(1 \leq i \leq k)$, if ($\mathscr{C}^i_{[\tau_j, \tau_{j+1}]}$) $= \mathscr{C}^c_{\mathscr{S}_{\mathscr{T}}}$ then insert the new $node^{new}_{TC} = (\mathscr{C}^i_{[\tau_j, \tau_{j+1}]}, [\tau_j, \tau_{j+1}])$ into left sub-tree of $node^{new}_{TC}$, and into the tail of L.

As we already pointed out, L maintains all current $\mathscr{C}^c_{\mathscr{S}}$. In order to check whether a trajectory cluster is closed or not, we only need to perform a backward closure checking by comparing with $node_{TC} \in L$ whose begin time is less than τ_{begin}.

Continuous query processing should manage continuous queries based on any mobile object and at any time. A threshold-based trajectory cluster (TTC) query involves three parameters: mobile object id ($\mathscr{O}_{id}$), time instant (τ) and minimum number of mobile objects ($min_{\mathscr{O}}$) in cluster. The Algorithm 8 shows pseudo code of TTC, based on DFS[12] algorithm for searching tree data structures.

---

[12]Depth-First-Search.

Algorithm 7 TraCluUpdate—Update Trajectory Clusters

Input: $\mathscr{C}_{\mathscr{L}_{[\tau_j, \tau_{j+1}]}} = \{\mathscr{C}^1_{[\tau_j, \tau_{j+1}]}, \mathscr{C}^2_{[\tau_j, \tau_{j+1}]}, \ldots, \mathscr{C}^k_{[\tau_j, \tau_{j+1}]}\}$, ▷existing line clusters
   $TC - tree$,   ▷a LCRS binary tree with multiple root nodes
   L,   ▷list of current closed trajectory clusters
   min_{ls} ▷minimum number of line segments in line-segment-cluster
   new node ← *L.head*
   while *node.next* ≠ *null* **do**
     new tcNode ← *node.next*
     new nextNode ← *tcNode.next*
     for all $\mathscr{C} \in \mathscr{C}_{\mathscr{L}_{[\tau_{i-1}, \tau_i]}}$ **do**
       if $\mathscr{C} \supseteq tcNode.\mathscr{O}_m$ **then**   ▷Rule R_2
         tcNode.end ← τ_{j+1}
       else if $\|\mathscr{C} \cap tcNode.\mathscr{O}_m\| \geq min_{ls}$ **then**   ▷Rule R_3
         new n ← $(\mathscr{C} \cap tcNode.\mathscr{O}_m, \tau_{j+1})$
         insertTCnode(tcNode, n)
         n.next ← *tcNode.next*
         tcNode.next ← *n*
       end if
     end for
     if $tcNode.child_{left}$ ≠ *null* **then**
       node.next ← *tcNode.next*
       tcNode.next ← *null*
     end if
     node ← *nextNode*
   end while
   for all $\mathscr{C} \in \mathscr{C}_{\mathscr{L}_{[\tau_{k-1}, \tau_k]}}$ **do**   ▷Rule R_4
     new n ← $(\mathscr{C}, [\tau_{k-1}, \tau_k])$
     if *backwardClosureCheck(null, n)* **then**
       insert n *into* $tcNode.childTree_{right}$ ▷Rule R_1
       node.next ← n
       node ← *node.next*
     end if
   end for
   insertNode(tcNode, n)
   if *backwardClosureCheck(tcNode, n)* **then**
     insert n *into* $tcNode.childTree_{left}$ ▷Rule R_1
   end if
   backwardClosureCheck(tcNode.root, n)
   for all ∈ *L.head* **do**
     if $node.\mathscr{O}_m = n.\mathscr{O}_m$ **then**
       return *false*
     end if
   end for
   return *true*

Algorithm 8 TTCQ—Threshold-based Trajectory Cluster Query

Input: $TC - tree$,   ▷a LCRS binary tree with multiple root nodes
          $\mathscr{O}_{id}$,   ▷id of mobile object
          τ,   ▷time instant
          $min_{\mathscr{O}}$ ▷minimum number of mobile objects
for all $root \in TC - tree$ **do**
    DFSQUERY($root.child_{left}$, $\mathscr{O}_{id}$, τ, $min_{\mathscr{O}}$)
end for
return $TTCs$
procedure DFSQUERY($node$, $\mathscr{O}_{id}$, τ, $min_{\mathscr{O}}$)
    if $node \neq null \wedge \tau \geq node.\tau_{begin}$ **then**
        if $\mathscr{O}_{id} \notin node.\mathscr{O}$ **then**
            DFSQUERY($node.child_{right}$, $\mathscr{O}_{id}$, τ, $min_{\mathscr{O}}$)
        else
            if $node.child_{left} \neq null \wedge \mathscr{O}_{id} \in node.child_{left}.\mathscr{O}$ **then**
                DFSQUERY($node.child_{left}$, $\mathscr{O}_{id}$, τ, $min_{\mathscr{O}}$)
            else
                $TTCs \leftarrow TTCs \oplus node$ ▷add node into TTCs
            end if
        end if
    end if
end procedure

4.3.3 Spatial Quincunx Lattices Based Clustering

The basic presumption of this method [13] is that a trajectory has a natural inter-
pretation as a discrete-time signal in which each trajectory point corresponds to an
impulse. The similarity between trajectories, i.e. trajectory clustering can be implemented by non separable transforms based on Fourier transformation of trajectories
in two-dimensional space. Non separable transforms allow us to deal with the whole
trajectory taking into account both dimensions and avoiding any approximation due
to mono-dimensional transform composition.

A prerequisite for mathematical transformation is the definition of the basic function and the characteristics of the trajectory search space. A *quincunx lattice* [39] is
the best representation for the trajectory search space.

Firstly, we provide the necessary background on polynomial algebras and their
connection to signal transforms in one and two dimensions. Therefore, we recall the
definition and some properties of Chebyshev polynomials, which are necessary to
algebraically describe spatial transforms.

4.3.3.1 Polynomial Algebra and Transforms

Chebyshev polynomials in one variable are defined by the recursion equation

$$Cn(x) = 2xC_{n-1}(x) - C_{n-2}(x), n \geq 2 \tag{4.26}$$

The exact form of C_n is determined by the initial conditions C_0 and C_1, which are chosen as polynomials of degree 0 and 1, respectively. The Eq. 4.26 implies that C_n is a polynomial of degree n for $n \geq 0$. The Chebyshev polynomials of the first kind, denoted with $C = T$ arise from

$$T_0(x) = 1, \quad T_1(x) = x \tag{4.27}$$

The Chebyshev polynomials of the third kind, denoted with $C = V$ arise from

$$V_0(x) = 1, \quad V_1(x) = 2x - 1 \tag{4.28}$$

Running the recurrence 4.26 in the other direction yields, for given initial conditions, Chebyshev polynomials C_n for negative indices n, and thus $\forall n \in \mathbb{Z}$.

We will use the following properties:

$$closed\ form : T_n = \cos n\theta, \quad x = \cos\theta; \tag{4.29}$$

$$V_n = \cos\frac{(n + \frac{1}{2})\theta}{\cos\frac{1}{2}\theta}, \quad x = \cos\theta \tag{4.30}$$

$$product : T_k C_n = (C_{n+k} + C_{n-k})/2, \quad C = T, V \tag{4.31}$$

$$relation : T_n + T_{n+1} = (x + 1)V_n \tag{4.32}$$

$$decomposition : T_{km} = T_k(T_m) \tag{4.33}$$

$$n\ zeros\ of\ T_n : \alpha_k = \cos(k + 1/2)\pi/n, \quad 0 \leq k < n \tag{4.34}$$

Definition 4.14 (*polynomial algebra—one variable*) An algebra A is a vector space closed under multiplication and the distributivity law holds. Examples include the set of complex numbers $\mathscr{A} = \mathbb{C}$ and the set of polynomials $\mathscr{A} = \mathbb{C}[x]$ in one variable, or in several variables $\mathscr{A} = \mathbb{C}[x_1, \ldots, x_k]$.

A polynomial algebra in one variable is the set of polynomials of degree less than $deg(p)$ with addition and multiplication modulo p:

$$\mathscr{A} = \mathbb{C}[x]/p(x) = \{q(x)|deg(q) < deg(p)\} \tag{4.35}$$

As a vector space, $\mathscr{A}$ has dimension $dim(\mathscr{A}) = deg(p)$. Further, $\mathscr{A}$ is obviously generated by x, since every element of A is a polynomial in x.

Definition 4.15 (*polynomial algebra—two variables*) A polygonal algebra in two variables[13] is defined as:

$$\mathcal{A} = \mathbb{C}[x, y]/\langle p(x, y), q(x, y)\rangle \tag{4.36}$$

Definition 4.16 (*polynomial transform—two variables*) To define a polynomial transform for the algebra in 4.36 we assume that the total degrees[14] of p and q are n and m, respectively, and that the equations $p(x, y) = q(x, y) = 0$ have precisely mn solutions $\alpha = (\alpha_k, \beta_k)_{0 \leq k < mn}$. The isomorphic decomposition of $\mathcal{A}$ into its spectrum is defined as:

$$\Phi : \mathcal{A} \rightarrow \bigoplus_{0 \leq k < mn} \mathbb{C}[x, y]/\langle x - \alpha_k, y - \beta_k\rangle \tag{4.37}$$

which implies that $dim(\mathcal{A}) = mn$. With respect to a basis $b = (p_l(x, y)|0 \leq l < mn)$ of $\mathcal{A}$, the polynomial transform for $\mathcal{A}$ is given by the matrix

$$\mathcal{P}_{b,\alpha} = [p_l(\alpha_k, \beta_k)]_{0 \leq k < mn} \tag{4.38}$$

The matrix $\mathcal{P}_{b,\alpha}$ is *polynomial transform* or *Fourier transform* transform for $\mathcal{A}$.

4.3.3.2 Spatial Quincunx Lattice and Transform

The two-dimensional transform $DCT - 3_n \otimes DCT - 3_n$ is[15] a polynomial transform for

$$\mathcal{A} = \mathbb{C}[x, y]/\langle T_n(x), T_n(y)\rangle \tag{4.39}$$

with basis

$$b = (T_i(x)T_j(y)|0 \leq i, j < n) \tag{4.40}$$

Assuming that n is even, and omitting the basis polynomials in 4.40 that do not reside on the quincunx lattice in Fig. 4.16 leaves exactly:

$$b = (T_{i,j}(x, y) = T_i(x)T_j(y)|i + j \equiv 0 \ mod \ 2) \tag{4.41}$$

---

[13]In contrast to the one-dimensional case, were one polynomial suffices, we need to compute modulo two or more polynomials to ensure that the dimension of $\mathcal{A}$ is finite.

[14]The total degree of $p(x, y)$ is defined as the maximum value of $i + j$, as i and j range over all summands $cx^i y^j of p(x, y)$.

[15]Discrete Cosine Transformation.

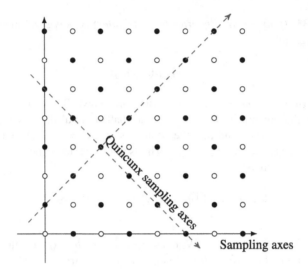

Fig. 4.16 A Quincunx lattice

Obviously, b spans a subvector space $\mathcal{B}$ of $\mathcal{A}$ of dimension $n^2/2$, and $\mathcal{B}$ is a subalgebra $\mathcal{B} < \mathcal{A}$ corresponding to the quincunx lattice in Fig. 4.16. To bring $\mathcal{B}$ into the from of 4.36 we need to find a set of generators, which then become our *variables*. Natural choices are

$$u = T_2(x) = 2x^2 - 1 \tag{4.42}$$

$$v = T_2(y) = 2y^2 - 1 \tag{4.43}$$

but their spanned algebra does not contain

$$w = T1(x)T1(y) = xy; \tag{4.44}$$

which thus we choose as the third generator. Since we have three polynomials u, v, w in two variables, they are dependent and they satisfy the polynomial relation

$$4w^2 - (u+1)(v+1) = 0 \tag{4.45}$$

Further, from 4.33, $T_{n/2}(u) = T_n(x) = 0$ in $\mathcal{A}$, and $T_{n/2}(v) = 0$. Thus, we obtain the polynomial algebra

$$\mathcal{B} = \mathbb{C}[u, v, w]/\langle T_{n/2}(u), T_{n/2}(v), 4w^2 - (u+1)(v+1)\rangle \tag{4.46}$$

Variable u acts on the quincunx lattice horizontally, whereas variable v acts vertically. The action of w, using 4.31, is given by

$$wT_{i,j} = (T_{i-1,j-1} + T_{i-1,j+1} + T_{i+1,j-1} + T_{i+1,j+1})/4 \qquad (4.47)$$

To compute the Fourier transform for $\mathscr{B}$, we need to express the basis b in 4.41 in u, v, w and compute the spectrum of $\mathscr{B}$, i.e., the solutions of

$$T_{n/2}(u) = T_{n/2}(v) = 4w^2 - (u+1)(v+1) = 0 \qquad (4.48)$$

Using 4.34, we get the $n^2/2$ solutions which define the spatial signal spectrum of $\mathscr{B}$:

$$u_k = \cos \frac{k+1/2}{n/2}\pi, \quad 0 \le k < n/2 \qquad (4.49)$$

$$v_l = \cos \frac{l+1/2}{n/2}\pi, \quad 0 \le l < n/2 \qquad (4.50)$$

$$w_{k,l,\pm} = \pm \frac{1}{2}\sqrt{(1+u_k) + (1+v_l)} \qquad (4.51)$$

4.3.3.3 Spatial Quincunx Distance

We assume that each trajectory point is an impulse in the overall signal associated to the trajectory. Therefore, we define the distance between two trajectories $\mathscr{T}_1$ and $\mathscr{T}_2$ by considering their their spectra $\mathscr{Q}(\mathscr{T}_1)$ and $\mathscr{Q}(\mathscr{T}_2)$, respectively. We compute their angular distances $d_{\widehat{\alpha\beta}}$ and $d_{\widehat{\gamma\delta}}$ as:

$$d_{\widehat{\alpha\beta}} = \arccos(u_{1k}) - \arccos(u_{2k}) \qquad (4.52)$$

$$d_{\widehat{\gamma\delta}} = \arccos(v_{1l}) - \arccos(v_{2l}) \qquad (4.53)$$

For each pair of $w_{k,l,\pm}$ we compute the modulo distance as

$$d_w = \sqrt{w^2_{1k,l,\pm} - w^2_{2k,l,\pm}} \qquad (4.54)$$

Finally, we define the overall quincunx lattice based distance between two trajectories T_1 and T_2 as:

$$\mathscr{Q}_{QL} = \sqrt{\sum_{k=1}^{n^2} d_w(k)^2 \cdot \cos(\mu(d_{\widehat{\alpha\beta}}) - \mu(d_{\widehat{\gamma\delta}}))} \qquad (4.55)$$

where $\mu(d_{\widehat{\alpha\beta}})$ is the average angle distance between each pair of u_k and $\mu(d_{\widehat{\gamma\delta}})$ is the mean between each pair of v_l.

This distance is capable to catch dissimilarity between T_1 and T_2 since it considers the difference in angular distances between them (the cos argument) while taking into account the overall extension of the spatial signal (the modulo distance). An interesting characteristic of 4.55 is that defined function is also a metric distance.

4.3.3.4 Incremental Clustering

Since we compute transforms using incremental approach [36], as new data stream arrives in a window, we only update transforms thus avoiding a complete recomputing of the signals, transforms and distances. This feature is very interesting and effective when dealing with trajectory streams.

Definition 4.17 (*Trajectory cluster*) Given a set of trajectories $\mathcal{T}$, a cluster $\mathscr{C}_{\mathcal{T}}$ is a subset $\mathscr{C}_{\mathcal{T}} \subseteq \mathcal{T}$ such that the distance between each pair of trajectories $\mathcal{T}_i \subseteq \mathscr{C}_{\mathcal{T}}$ and $\mathcal{T}_j \subseteq \mathscr{C}_{\mathcal{T}}$ is minimum, and the distance between each pair of trajectories $\mathcal{T}_i \subseteq \mathscr{C}_{\mathcal{T}}$ and $\mathcal{T}_j \not\subseteq \mathscr{C}_{\mathcal{T}}$ is maximized with regard to the chosen metric.

The Algorithm 9 is adapted for clustering evolving trajectory data streams. The data set being mined is defined as a sliding window over the continuous spatio-temporal data stream. As a new window slide $\mathscr{S}_{new}$ is loaded, the clustering is performed by invoking *newCluster* that implements *k-means++* algorithm [6].

Algorithm 9 Fourier$_{2D}$ Incremental Clustering

Input: $\mathcal{T}_{\mathscr{S}}$ ▷trajectory stream
Vars: $\mathscr{S}_{new}$,   ▷new slide of the input trajectories
         $\mathscr{S}_{exp}$,   ▷expiring slide of the input trajectories
         $\mathscr{C}_{new}$ ▷new cluster
  $\mathscr{C}_{\mathcal{T}} \leftarrow \varnothing$
  for all $\mathscr{S}_{new}$ **do**
      $\mathscr{C}_{new} \leftarrow newCluster(\mathcal{T}_{\mathscr{S}})$
      $mergeClusters(\mathscr{C}_{new}, \mathscr{C}_{\mathcal{T}})$
      for all $\mathcal{T} \in \mathcal{T}_{\mathscr{S}}$ **do**
        $annotateTrajectory(\mathcal{T})$
      end for
  end for
  for all $\mathscr{S}_{exp}$ **do**
    $discardOldestTrajectories(\mathscr{C}_{\mathcal{T}})$
    $compactClusters(\mathscr{C}_{\mathcal{T}})$
  end for
  return $\mathscr{C}_{\mathcal{T}}$

The *mergeClusters* function keeps fresh clusters and eliminates possible concept drift. If the cluster assignment for trajectories belonging to new slide $\mathscr{S}_i$ conforms to the current one, the new current cluster centers are (eventually) recomputed. Otherwise, if a new cluster is generated, the cluster data are updated in order to take them into account for next slides. This process is driven by minimizing $\mathcal{D}_{QL}$ function. Given the centers of the current clusters, the center of the each cluster is updated using the trajectory that minimizes the $\mathcal{D}_{QL}$ with regard to the new set of trajectories belonging to the cluster. The result of this step is continuously updating of a cluster assignment, therefore it results impervious to the eventual concept drift. To allow successive delta maintenance, the trajectories in the current window slide are annotated by their timestamps. As a window slide expires, the oldest trajectories are discarded and the clusters that could become loosely compact after trajectory deletions are re-compacted. The function *compactClusters* recomputes cluster centers with regard to $\mathcal{D}_{QL}$ minimization as in *mergeClusters*. Therefore, the clusters are recomputed including the new trajectories and assigning the new cluster centers. It is worth to note that no threshold parameter is provided since $\mathcal{D}_{QL}$ is metric, and thus it preserves the fundamental *k-means++* features.

4.4   Bibliographic Notes

The literature on spatio-temporal, and on mobility data in particular, is rather extensive and heterogeneous. In the following, we provide a list of essential bibliographic references including those describing fundamental methods and algorithms discussed in this chapter.

Several global distance functions were introduced: geographic distance in [33], Dynamic Time Warping (DTW) in [47], LCSS (Longest Common SubSequence Measure) in [43] and ERP (Edit distance with Real Penalty) in [12]. Similarly, a number of local distance functions were introduced: trajectory-Hausdorff distance in [30], MBR (Minimum Bounding Rectangle) in [2], MBB (Minimum Bounding Box) in [16], spatio-temporal similarity measure based on Minimum Boundix Box (MBB) was introduced in [50], trajectory-segment distance in [27] and CPA (Closest Point of Approach) in [7]. An overview of trajectory distance measures was also given in [25].

Data stream clustering is an active research area and several algorithms have been proposed to perform unsupervised learning. Silva et. al. [15] present a survey of data stream clustering algorithms, providing a thorough discussion of the main features of state-of-the-art algorithms. These clustering algorithms handles only incremental data, and therefore they are not directly applicable to trajectories embedded into spatio-temporal data streams.

Currently, the number of works in this very young and challenging discipline is rather limited. Most notable among early works is that of [24], in which authors proposed moving-object clustering. The work presented in [40] investigates the problem of discovering *traveling companions* (i.e., object groups that travel together) from trajectory streams. To facilitate scalable and flexible data processing, the authors introduced data structure called *travelling buddy*.

A very detailed description of spatio-temporal data clustering methods and algorithms described in this chapter, namely TCMM (Trajectory Clustering using Micro- and Macro- clustering), CTraStream, and incremental clustering by exploiting Fourier transformation for spatial quincunx lattices based clustering, are presented in [32], [46, 48] and [13] respectively. These references also contain extensive performance evaluations and comparisons.

References

1. Aggarwal, C.C., Han, J., Wang, J., Yu, P.S.: A framework for clustering evolving data streams. In: VLDB. pp. 81–92 (2003). http://www.vldb.org/conf/2003/papers/S04P02.pdf
2. Anagnostopoulos, A., Vlachos, M., Hadjieleftheriou, M., Keogh, E.J., Yu, P.S.: Global distance-based segmentation of trajectories. In: Eliassi-Rad, T., Ungar, L.H., Craven, M., Gunopulos, D. (eds.) Proceedings of the Twelfth ACM SIGKDD International Conference on Knowledge Discovery and Data Mining, Philadelphia, PA, USA, August 20–23. pp. 34–43. ACM, New York (2006). http://doi.acm.org/10.1145/1150402.1150411
3. Andersson, M., Gudmundsson, J., Laube, P., Wolle, T.: Reporting leaders and followers among trajectories of moving point objects. GeoInformatica 12(4), 497–528 (2008). http://dx.doi.org/10.1007/s10707-007-0037-9
4. Andrienko, G.L., Andrienko, N.V., Bak, P., Keim, D.A., Kisilevich, S., Wrobel, S.: A conceptual framework and taxonomy of techniques for analyzing movement. J. Vis. Lang. Comput. 22(3), 213–232 (2011). http://dx.doi.org/10.1016/j.jvlc.2011.02.003
5. Ankerst, M., Breunig, M.M., Kriegel, H., Sander, J.: OPTICS: ordering points to identify the clustering structure. In: Delis, A., Faloutsos, C., Ghandeharizadeh, S. (eds.) SIGMOD 1999, Proceedings ACM SIGMOD International Conference on Management of Data, June 1–3, 1999, Philadelphia, Pennsylvania, USA. pp. 49–60. ACM Press, New York (1999). http://doi.acm.org/10.1145/304182.304187
6. Ankerst, M., Breunig, M.M., Kriegel, H., Sander, J.: OPTICS: ordering points to identify the clustering structure. In: Delis, A., Faloutsos, C., Ghandeharizadeh, S. (eds.) SIGMOD 1999, Proceedings ACM SIGMOD International Conference on Management of Data, June 1–3, 1999, Philadelphia, Pennsylvania, USA. pp. 49–60. ACM Press, New York (1999). http://doi.acm.org/10.1145/304182.304187
7. Arumugam, S., Jermaine, C.: Closest-point-of-approach join for moving object histories. In: Liu, L., Reuter, A., Whang, K., Zhang, J. (eds.) Proceedings of the 22nd International Conference on Data Engineering, ICDE 2006, 3–8 April 2006, Atlanta, GA, USA. p. 86. IEEE Computer Society, New York (2006). http://dx.doi.org/10.1109/ICDE.2006.36
8. Bahmani, B., Moseley, B., Vattani, A., Kumar, R., Vassilvitskii, S.: Scalable K-Means++. PVLDB 5(7), 622–633 (2012)
9. Benkert, M., Gudmundsson, J., Hübner, F., Wolle, T.: Reporting flock patterns. Comput. Geometry 41(3), 111–125 (2008)
10. Cao, F., Ester, M., Qian, W., Zhou, A.: Density-based clustering over an evolving data stream with noise. In: Ghosh, J., Lambert, D., Skillicorn, D.B., Srivastava, J. (eds.) Proceedings of the Sixth SIAM International Conference on Data Mining, April 20–22, 2006, Bethesda, MD, USA. pp. 328–339. SIAM (2006). http://dx.doi.org/10.1137/1.9781611972764.29
11. Carneiro, C., Alp, A., de Macêdo, J.A.F., Spaccapietra, S.: Advanced data mining method for discovering regions and trajectories of moving objects: "ciconia ciconia" scenario. In: Bernard, L., Friis-Christensen, A., Pundt, H. (eds.) The European Information Society: Taking Geoinformation Science One Step Further, Proceedings of the 11th AGILE Conference, Lecture Notes in Geoinformation and Cartography, Girona, Spain, 5–8 May, 2008. pp. 201–224. Springer, Heidelberg (2008). http://dx.doi.org/10.1007/978-3-540-78946-8_11

12. Chen, L., Ng, R.T.: On the marriage of lp-norms and edit distance. In: Nascimento, M.A., Özsu, M.T., Kossmann, D., Miller, R.J., Blakeley, J.A., Schiefer, K.B. (eds.) (e)Proceedings of the Thirtieth International Conference on Very Large Data Bases, Toronto, Canada, August 31–September 3 2004. pp. 792–803. Morgan Kaufmann (2004). http://www.vldb.org/conf/2004/RS21P2.PDF
13. Costa, G., Manco, G., Masciari, E.: Dealing with trajectory streams by clustering and mathematical transforms. J. Intell. Inf. Syst. 42(1), 155–177 (2014). http://dx.doi.org/10.1007/s10844-013-0267-2
14. Dodge, S., Weibel, R., Lautenschütz, A.K.: Towards a taxonomy of movement patterns. Inf. Vis. 7(3–4), 240–252 (2008)
15. de Andrade Silva, J., Faria, E.R., Barros, R.C., Hruschka, E.R., de Carvalho, A.C.P.L.F., Gama, J.: Data stream clustering: A survey. ACM Comput. Surv. 46(1), 13 (2013). http://doi.acm.org/10.1145/2522968.2522981
16. Elnekave, S., Last, M., Maimon, O., Ben-Shimol, Y., Einsiedler, H.J., Friedman, M., Siebert, M.: Discovering regular groups of mobile objects using incremental clustering. In: 5th Workshop on Positioning, Navigation and Communication, WPNC 2008, Leibniz Universität Hannover, Hannover, Germany, March 27, 2008. pp. 197–205. IEEE, New York (2008). http://dx.doi.org/10.1109/WPNC.2008.4510375
17. Ester, M., Kriegel, H., Sander, J., Xu, X.: A density-based algorithm for discovering clusters in large spatial databases with noise. In: Simoudis, E., Han, J., Fayyad, U.M. (eds.) Proceedings of the Second International Conference on Knowledge Discovery and Data Mining (KDD-96), Portland, Oregon, USA. pp. 226–231. AAAI Press, Massachusetts (1996). http://www.aaai.org/Library/KDD/1996/kdd96-037.php
18. Folino, G., Spezzano, G.: SPARROW: A spatial clustering algorithm using swarm intelligence. In: Hamza, M.H. (ed.) The 21st IASTED International Multi-Conference on Applied Informatics (AI 2003), February 10–13, 2003, Innsbruck, Austria. pp. 50–55. IASTED/ACTA Press, Canada (2003)
19. Gama, J.: Knowledge Discovery from Data Streams. Chapman and Hall/CRC Data Mining and Knowledge Discovery Series, CRC Press, Florida (2010). http://www.crcpress.com/product/isbn/9781439826119
20. Giannotti, F., Nanni, M., Pedreschi, D., Pinelli, F., Renso, C., Rinzivillo, S., Trasarti, R.: Unveiling the complexity of human mobility by querying and mining massive trajectory data. VLDB J. 20(5), 695–719 (2011). http://dx.doi.org/10.1007/s00778-011-0244-8
21. Grosan, C., Abraham, A.: Intelligent Systems: A Modern Approach, 1st edn. Springer Publishing Company, Incorporated, Heidelberg (2011)
22. Han, J., Lee, J.G., Kamber, M.: An overview of clustering methods in geographic data analysis. In: Miller, H.J., Han, J. (eds.) Geographic Data Mining and Knowledge Discovery, 2nd edn. pp. 149–187. CRC Press, Boca Raton (2009)
23. Han, J., Kamber, M., Pei, J.: Data Mining: Concepts and Techniques, 3rd edn. Morgan Kaufmann Publishers Inc., San Francisco (2011)
24. Jensen, C.S., Lin, D., Ooi, B.C.: Continuous clustering of moving objects. IEEE Trans. Knowl. Data Eng. 19(9), 1161–1174 (2007). http://dx.doi.org/10.1109/TKDE.2007.1054
25. Jeung, H., Yiu, M.L., Jensen, C.S.: Trajectory pattern mining. In: Zheng, Y., Zhou, X. (eds.) Computing with Spatial Trajectories, pp. 143–177. Springer, Heidelberg (2011). http://dx.doi.org/10.1007/978-1-4614-1629-6
26. Jeung, H., Yiu, M.L., Zhou, X., Jensen, C.S., Shen, H.T.: Discovery of convoys in trajectory databases. PVLDB 1(1), 1068–1080 (2008). http://www.vldb.org/pvldb/1/1453971.pdf
27. Jeung, H., Yiu, M.L., Zhou, X., Jensen, C.S., Shen, H.T.: Discovery of convoys in trajectory databases. CoRR (2010). http://arxiv.org/abs/1002.0963
28. Kalnis, P., Mamoulis, N., Bakiras, S.: On discovering moving clusters in spatio-temporal data. In: Medeiros, C.B., Egenhofer, M.J., Bertino, E. (eds.) Advances in Spatial and Temporal Databases, 9th International Symposium, SSTD 2005, Angra dos Reis, Brazil, August 22–24, 2005, Proceedings. Lecture Notes in Computer Science, vol. 3633, pp. 364–381. Springer (2005). http://dx.doi.org/10.1007/11535331_21

29. Kranen, P., Assent, I., Baldauf, C., Seidl, T.: The ClusTree: indexing micro-clusters for anytime stream mining. Knowl. Inf. Syst. **29**(2), 249–272 (2011). http://dx.doi.org/10.1007/s10115-010-0342-8

30. Lee, J., Han, J., Whang, K.: Trajectory clustering: a partition-and-group framework. In: Chan, C.Y., Ooi, B.C., Zhou, A. (eds.) Proceedings of the ACM SIGMOD International Conference on Management of Data, Beijing, China, June 12–14, 2007. pp. 593–604. ACM (2007). http://doi.acm.org/10.1145/1247480.1247546

31. Li, Z., Han, J., Ji, M., Tang, L.A., Yu, Y., Ding, B., Lee, J., Kays, R.: MoveMine: Mining moving object data for discovery of animal movement patterns. ACM TIST **2**(4), 37 (2011). http://doi.acm.org/10.1145/1989734.1989741

32. Li, Z., Lee, J., Li, X., Han, J.: Incremental clustering for trajectories. In: Kitagawa, H., Ishikawa, Y., Li, Q., Watanabe, C. (eds.) Database Systems for Advanced Applications, 15th International Conference, DASFAA 2010, Tsukuba, Japan, April 1–4, 2010, Proceedings, Part II. Lecture Notes in Computer Science, vol. 5982, pp. 32–46. Springer (2010). http://dx.doi.org/10.1007/978-3-642-12098-5_3

33. Liu, H., Schneider, M.: Similarity measurement of moving object trajectories. In: Proceedings of the Third ACM SIGSPATIAL International Workshop on GeoStreaming. pp. 19–22. IWGS '12, ACM, New York, NY, USA (2012). http://doi.acm.org/10.1145/2442968.2442971

34. Lloyd, S.P.: Least squares quantization in PCM. IEEE Trans. Inf. Theory **28**(2), 129–136 (1982). http://dx.doi.org/10.1109/TIT.1982.1056489

35. Nanni, M., Pedreschi, D.: Time-focused clustering of trajectories of moving objects. J. Intell. Inf. Syst. **27**(3), 267–289 (2006). http://dx.doi.org/10.1007/s10844-006-9953-7

36. Oppenheim, A.V., Schafer, R.W.: Discrete-time Signal Processing, 3rd edn. Pearson Education Limited, Essex, UK (2009)

37. Pelekis, N., Kopanakis, I., Kotsifakos, E.E., Frentzos, E., Theodoridis, Y.: Clustering trajectories of moving objects in an uncertain world. In: Wang, W., Kargupta, H., Ranka, S., Yu, P.S., Wu, X. (eds.) ICDM 2009, The Ninth IEEE International Conference on Data Mining, Miami, Florida, USA, 6–9 December 2009. pp. 417–427. IEEE Computer Society (2009). http://dx.doi.org/10.1109/ICDM.2009.57

38. Pelekis, N., Theodoridis, Y.: Mobility Data Management and Exploration. Springer (2014). http://dx.doi.org/10.1007/978-1-4939-0392-4

39. Püschel, M., Rötteler, M.: Fourier transform for the spatial quincunx lattice. In: Proceedings of the 2005 International Conference on Image Processing, ICIP 2005, Genoa, Italy, September 11–14, 2005. pp. 494–497. IEEE (2005). http://dx.doi.org/10.1109/ICIP.2005.1530100

40. Tang, L.A., Zheng, Y., Yuan, J., Han, J., Leung, A., Peng, W., Porta, T.F.L.: A framework of traveling companion discovery on trajectory data streams. ACM TIST **5**(1), 3 (2013). http://doi.acm.org/10.1145/2542182.2542185

41. Theodoridis, S., Koutroumbas, K.: Pattern Recognition, 4th edn. Academic Press, Burlington, MA, USA (2008)

42. Vieira, M.R., Tsotras, V.J.: Spatio-Temporal Databases—Complex Motion Pattern Queries. Springer Briefs in Computer Science, Springer, Heidelberg (2013). http://dx.doi.org/10.1007/978-3-319-02408-0

43. Vlachos, M., Gunopulos, D., Kollios, G.: Discovering similar multidimensional trajectories. In: Agrawal, R., Dittrich, K.R. (eds.) Proceedings of the 18th International Conference on Data Engineering, San Jose, CA, USA, February 26—March 1, 2002. pp. 673–684. IEEE Computer Society (2002). http://dx.doi.org/10.1109/ICDE.2002.994784

44. Wang, W., Yang, J., Muntz, R.R.: STING: A statistical information grid approach to spatial data mining. In: Jarke, M., Carey, M.J., Dittrich, K.R., Lochovsky, F.H., Loucopoulos, P., Jeusfeld, M.A. (eds.) VLDB'97, Proceedings of 23rd International Conference on Very Large Data Bases, August 25–29, 1997, Athens, Greece. pp. 186–195. Morgan Kaufmann (1997). http://www.vldb.org/conf/1997/P186.PDF

45. Wu, F., Lei, T.K.H., Li, Z., Han, J.: MoveMine 2.0: Mining object relationships from movement data. PVLDB **7**(13), 1613–1616 (2014). http://www.vldb.org/pvldb/vol7/p1613-wu.pdf

46. Yanwei, Y., Qin, W., Xiaodong, W.: Continuous clustering trajectory stream of moving objects. Commun. China **10**(9), 120–129 (2013). http://ieeexplore.ieee.org/xpl/articleDetails. jsp?arnumber=6623510

47. Yi, B., Jagadish, H.V., Faloutsos, C.: Efficient retrieval of similar time sequences under time warping. In: Urban, S.D., Bertino, E. (eds.) Proceedings of the Fourteenth International Conference on Data Engineering, Orlando, Florida, USA, February 23–27, 1998. pp. 201–208. IEEE Computer Society (1998). http://dx.doi.org/10.1109/ICDE.1998.655778

48. Yu, Y., Wang, Q., Wang, X., Wang, H., He, J.: Online clustering for trajectory data stream of moving objects. Comput. Sci. Inf. Syst. **10**(3), 1293–1317 (2013). http://dx.doi.org/10.2298/ CSIS120723049Y

49. Zhang, T., Ramakrishnan, R., Livny, M.: BIRCH: an efficient data clustering method for very large databases. In: Jagadish, H.V., Mumick, I.S. (eds.) Proceedings of the 1996 ACM SIGMOD International Conference on Management of Data, Montreal, Quebec, Canada, June 4–6, 1996. pp. 103–114. ACM Press (1996). http://doi.acm.org/10.1145/233269.233324

50. Zhao, X., Xu, W.: A new measurement method to calculate similarity of moving object spatio-temporal trajectories by compact representation. Int. J. Comput. Intell. Syst. **4**(6), 1140–1147 (2011). http://dx.doi.org/10.1080/18756891.2011.9727862

References

120

Index

© The Author(s) 2016
Z. Galić, *Spatio-Temporal Data Streams*, SpringerBriefs
in Computer Science, DOI 10.1007/978-1-4939-6575-5

Printed in the United States
By Bookmasters